# THE NEW PHYSICS

# OF MULTIDIMENSIONAL TIME

*(Hylomorphism: **The Theory of Everything**)*

Published by WisdomCS, LCC
Wisdom Cornerstone
4711 Hope Valley Rd. PMB220
Durham NC 27707-5651
http://www.wisdomcs.com



First edition
Printed in U.S.A
ISBN   978-0-9881800-2-4

# THE NEW PHYSICS

## OF MULTIDIMENSIONAL TIME

*(Hylomorphism: The Theory of Everything)*

*Naturam cognoscere, suffragatur Dei gloriae*

José Santiago Flores B.

This page is intentionally left in blank

a

## <u>Preface</u>

This publication presents a **new,** more complete version of the previously published, *"Physics of Multidimensional Time."* This current edition also offers a more detailed explanation on each subject, although the basic concepts remain the same; furthermore, some minor misstatements have been corrected.

Physics requires a unified theory explaining natural phenomena as observed in the micro and macro cosmos, all we have at this time, is a collection of three main proposals that together constitute an incomplete and disjointed theory; the first, the "Theory of Relativity" (TR) postulates a four-dimensional space-time continuum where material objects interact; this approach implicitly assumes that space is a physical entity, which sustains the environment by placement of matter. The second proposal, "Quantum Mechanics" (QM) investigates natural phenomena from a probabilistic point of view using mathematics and statistics as prospective and analytic tools evaluating the probabilities of cause and effect. A third proposal, the "Electromagnetic Theory" (ELM) studies space-time interactions between electric and magnetic fields. All three theories have proven valid in explaining phenomena in their individual fields; but they have been invaded by complicated mathematical theorems and procedures that, when applied by brilliant minds in the quest for intellectual pedagogy may end up identifying "truths" rather than "realities."

Physicists turned to mathematics as a lifesaver, using it as a prospective tool without considering the deepest connotation - mathematical results are truths always related to their axiomatic bases, if the fundamental supposition as stated in a formula were partially false, one would get "correct" unrealistic answers. Mathematics is a tool that facilitates finding answers related to a basic postulate; we must always keep in mind that the survival of the human species is the result of the interpretation and application of recursive logic when observing nature's "reality" rather than the blind application of "mathematical truths." When an observed causal relation is described in terms of a hypothesis, mathematical formulae can illustrate all the truths this hypothesis contains. When some of these "truths" do not match observation, the initial hypothesis must be revised; regardless of the lower "investigation productivity" that this approach may generate, as a pure logic protocol is less effective than applying mathematical theorems, and it has the great advantage of constantly questioning the validity of the initial hypothesis.

By adhering to paradigms established by genial theoreticians, current physical theories in some instances confound the essentials from the contingent, the causes from the effects, and the substances from the forms. For example, the currently accepted concept of speed of light is inaccurate; it is also universally postulated that lighter means smaller, which is not applicable to the micro cosmos; current theories judge that "red shift" is a consequence of gravity; that electrons are small particles orbiting the atomic nucleus; that flat space matrix is disturbed by mass and even in the most basic physics textbooks it is stated that magnetism is a primary physical entity, and we may name many other paradigms that are supporting current physics. These examples, and many others, which assume a prejudicial preference *this* versus *that*, result in laws, formulas, and mathematical models that may end up predicting, modeling, and detecting entelechies. Even "experimental reality" could suffer misinterpretation, if influenced by the same paradigms which support the theory, that the experiment attempts proving; for instance, TR exhibits the "Red Shift" effect as proof that gravity attracts light, while the Theory of Hylomorphism (TH) refers to the same fact as the evidence of the opposite.

This behavioral preference is not particular to physics; scientific knowledge has always been initially guided by paradigms; Plato's "Academia" entrance announced: "*Do not enter he who does not know geometry,*" postulating that geometry was the base for all scientific knowledge. Today, twenty-six centuries later, there are knowledgeable physicists' that would say, "*You will not teach physics if you do not trust the String Theory,"* which neglects the essential ingredients for scientific development: freedom, criticism, and imagination. Scientists and researchers must be experts with extraordinary capability in observing phenomena under new perspectives.

The "Theory of Hylomorphism" (TH) as presented herein, demonstrates that absolute universal time_is the cornerstone supporting the universe, whereas space, gravity, and electricity are subsidiary entities of time and matter.

From the perspective of TH, physics appears as a unified science while current theories are acknowledged as specialized methods guiding the remote observation and statistical analysis of natural phenomena. TH explores the fundamentals of physical sciences, and consequently demands a revision and redefinition of basic concepts such as space, time, speed, acceleration, and force; adding greater conceptual richness and precision to their classical definitions. TH develops from the premise that universal raw material has two components: Time that supports and imposes the principle of causality, and an ontological ingredient represented by a binary vector which, by arranging "Hyle" time structures, manifest itself as matter, energy, space, gravity, electric charge, and magnetism. The "Hylomorphos" denomination is adopted as a deserved tribute to the genius of Aristotle, who twenty-six centuries ago, proposed a universe composed of a prime matter (Hyle) that takes diverse forms (morphos) that erect everything.

The development of a physics theory includes: contemplation, analysis, comprehension, speculation, and proof predictions. A hypothesis not only has to "make sense," but also must prove its validity by being compatible with the concepts and equations from other theories that prove matching nature's behavior. In this sense, <u>TH is a theory that explains many experimental facts already experimentally proven</u>, for example, the origin of gravity, Lorentz's contraction, Planck's equation's, facts that sustain Schrodinger's conjecture, the universal law of gravitation, magnetism, 'c' as a universal constant, Pauli's exclusion principle, Einstein's equivalence principle, and the super speed of some receding galaxies among other observed phenomena.

It is probable that investigators, habituated in deducing cause-effect relationships: from sophisticated costly experiments, or by using complex mathematical modeling in super computers, may find TH excessively simple and highly speculative; however, when a speculation explains all physics phenomena so coherently, the methodology used in proposing it, or the proponent's titles and curriculum should be irrelevant.

The TH prognostics, consequences, proposals, and predictions include subjects such as, at the beginning of time, universe conception as "Let there be light," versus the Big-Bang theory; how and why gravity works; how an electric field is created; why black holes do not exist; limits of Newton's gravitational law; explanations of black matter effects; the source of black energy; discusses why future space ships may go faster than the speed of light; proposes the possibility canceling gravity; elucidate quantum behavior of particles, to name a few.

I am aware that the proposal, as herein presented may contain unintentional inaccuracies, mistaken proposals, errors, and other imperfections; I also recognize that TH is incomplete and it will remain as it, for some time. As it stands, this study represents only the starting point; the reader should recognize that the magnitude of the missing theoretical explanations exceeds any personal capability. The limited theoretical analysis is intended for showing and explaining basic concepts; current resources within my reach have not allowed working on detailed numerical analysis; priority has been given to render solid concepts, and proposing guidelines and ideas for future research. The text of this study often sacrifices proper semantics by maintaining conventional (perhaps inappropriate) names and denominations making subjects easier to understand. There are concepts in TH that are so far removed from conventionality, so they may appear excessively abstract and speculative. This book goal is to offer a new perspective to the scientific community.

Considering that the objective of this paper explains for the first time new concepts that unify distant branches of physics, present study cannot analyze every subject with all the depth it deserves before passing to another; the reader may perceive that subjects jump back-and-forth many times in a repetitive text; however, this was intentional in order to take advantage on the fact that reading some subjects for a second time provokes deeper understanding.

It will also may be strange not seeing tensor notation in the mathematical formulations, but most concepts are clearer when explained with vectors in a two dimensional environment. I have chosen to make comprehension easier, hoping, thereby, increase the number of novice students reading this work

The culture in theoretical physics investigation must evolve by staying closer to common sense than to mathematical speculations; TH is just a first step.

## A unifying theory

For almost a century theoretical physics devoted efforts finding what A. Einstein referred as the "Theory of everything," a unifying proposal able to explain contradictory aspects of physics theories.

This theory will establish a new way to perceive nature, and must explain the source of already known basic natural laws, while its proposals should be in accordance to experimental data. Considering that the previous paragraph describes the final objective of physics as a scientific discipline, it will not be easy to fulfill even partially this supreme objective. It will demand REVISING ALL CLASSICAL CONCEPTS UNDER A DIFFERENT PERSPECTIVE, and this problem will not be solved by simply applying matrix operators to already known variables, neither explaining everything in a couple of pages, nor based on classical dogmas.

It is understandable that such a quest will trek across unexplored routes, with the end goal of presenting an exceptionally original proposal. Furthermore, it is reasonable assuming that on the exploration will not accept blindly, existing paradigms, nor follow classic mathematical procedures which failed, at least partially, in explaining nature behavior, moreover, it is quite possible that such a theory will modify current basic definitions and fundamental concepts.

Consequently, it should not be a surprise that this proposal probably will be written in an unconventional way, not supported by classical arguments, and will not follow common mathematical procedures proving its assertions. All these factors added together, might make it difficult for the reader to accept it easily. The reader, then, is challenged to adop an open attitude towards analyzing arguments which may oppose lifelong believes. It may be difficult for human nature to freely think when analyzing dogmas that support his or her self-confidence and prestige academically, there may be uncomfortable to change the targets of investigations. History proves, unequivocally, as shown by Galileo's experience; repulsion is the destiny of all unconventional proposals before recognizing its importance as an important legacy to human knowledge.

What will be the characteristics of a new proposal that unifies physics?

- Does it must base on the same roots of current theories? The investigation target demands being free of prejudice.
- Will it be explained following the standard mathematical procedures? NOT necessarily.

- Will it observe current scientific papers protocols? NOT essential; its objective is proposing new visions, not necessarily adhere to pedagogic references.
- Will it ignore current "principles" and "laws" from the very beginning? Possibly, YES.
- Will it respect current interpretations on space - time relations? NOT necessarily.
- Will it base its analysis on dogmatic "principles" and "conjectures" of current theories? NO, on the contrary it should be able to explain their causes.
- Will it preserve current definitions and notation of fundamental variables? NOT necessarily.
- Will the person presenting it have diplomas from prestigious universities? NOT necessarily.

A new theory must be complementary to the outcomes achieved by classic theories: the Theory of Relativity, Quantum Mechanics and Electromagnetic Theory. It should be comprehensive by explaining the mechanics of gravity, electricity, atomic structure, beta decay, and so forth. It should explain and validate Newton's Laws, Ampere's Relation, the Einstein equations, and the Einstein-Planck relation and basic laws. Finally, this theory should explain its agreements and disagreements with current dogmatic principles regarding, wave-matter duality light speed limit, space curvature, quarks, orbital, black holes, dark matter and spatial expansion, among others.

## The Theory of Hylomorphism

The Theory of Hylomorphism (TH) proposes new solutions in this book which are so coherent and universal that is highly probable be the base for a "new physics." TH agrees with QM that the laws ruling interaction are statistically driven, but also demonstrates that phenomena between individual particles is always absolutely causal; TH also agrees with some of the TR postulates, but it explains clearly

h

that absolute time is necessary to orderly establish an interactive universe.

What I consider the most important achievements of TH is that it restores the concept that common sense that science should always exercise, and its proposal dismisses indeterminacies, dualities, relativities, and other misleading concepts which do not observe the two cornerstones of natural universe - substance and causality.

## TH proposals summary

The TH analyzes current foundations of physics by exposing another reality with a new perspective that emphasizes a model based on the physical interaction among individual particles, which is ultimately the primary source of all natural phenomena. The validity of this proposal is confirmed by the consistency of its predictions with experimental observations; thus, TH "rediscovers," various laws such as: Newton's Law; Coulomb's Law; Ampere's Law; Einstein-Planck Relation, among others. The fundamental postulate of this theory is that natural communication is provided by entities, which are made up with the same substance (Hyle) than the material particle, which emits it. The information carriers must be oscillating entities; otherwise, they would be unable storing usable information. The only way to read information is by detecting the relative phase shift between the received oscillation and a referential oscillation given by a master clock[1]. Therefore, the only natural basic parameter ruling everything is time.

The TH proposes that the universe resides in an environment of four continuous, isotropic and homogeneous time dimensions; one absolute time, and three travel-time (TT) dimensions. Therefore,

---

[1] Like in radio reception; one must have a reference oscillator to detect and decode: FM, PCM, or a radar signal.

"space" is a mental abstraction of the three dimensions of travel time, which are converted to the arbitrarily set distance units, by applying a conversion factor "c."

Every particle is a continuous one-dimensional Hyle structure travelling on a straight trajectory. The oscillation plane orientation with respect to the direction of displacement defines the type of particle. Every segment travels away from the emitter at the rate of one second in the travel time dimension per second of elapsed absolute time; however, there may be asymmetric structures where the spatial period is shorter or larger than the nominal value set by absolute time, affecting (temporarily) the isometric quality of the four time dimensions.

## Laws and parameters

Some of the physics laws, phenomena, and parameters analyzed in this book under TH perspective are:

Particle and photon geometric and elastic properties: frequency, mass, energy, Einstein-Planck relation, law of constant angular momentum, Spin, $E = m c^2$.

Communication: emission and absorption. The origin of: Gravity; Newton's law; Coulomb's law; magnetic field, Ampere's law; magnetic moment; beta emission.

Composite Hyle structures: Atomic geometry, nuclear geometry.

Cosmology: expansion of universe, black holes, dark matter, speed limit, the origins of universe.

j

This page is intentionally left in blank

i

# CONTENTS

Preface

## Chapter I
## Introductory synthesis of the proposal  1



## Chapter II
## The four time dimensions  33
### I.- Space



### II.- TIME

# ANNEX I
## Comparative review with current Theories A



# ANNEX II
## Table: New Natural System of Units    GG

This page is intentionally left in blank

# NEW PHYSICS OF MULTIDIMENSIONAL TIME

The Theory of Hylomorphism

## Chapter I

## Introductory synthesis of the proposal

Ultimate science objective aims fulfilling human aspiration for knowledge and comprehension on natural phenomena. The established objective demands discovering the underlying relation between cause and effect in the observed event under study. The particular goal for natural sciences is searching for the rules that will enable predicting results by proposing theories not contradicting basic logic principles; logic constitutes the foundation for science and mathematics, not the opposite.

The Theory of Hylomorphism proposal, aspires helping humankind going forward towards those targets, by providing a solution that unifies existing theories in physics; the TH follows an absolutely unconventional protocol since normally, a research starts when observing a phenomenon, exploring its cause and proposing a probable relations between variables; finally, one test if the predictions describe nature's behavior. Moreover, the "Theory of Hylomorphism" (TH) illustrates a novel and surprisingly simple perspective perceiving nature, by explaining most of the basic phenomena, including those that current theories do not enlighten achieving this goal; the TH follows a pure logic approach. It is important saying that this proposal does not pretend to solve all current mysteries in physics; obviously, I do not pretend having the time, capacity, or required knowledge to accomplish that unreachable target; nevertheless, I am confident that this proposal will sooner or later be the cornerstone for raising a solid structure for theoretical physics.

Currently we are in a privileged position reaching the unification of theories; we have access to a huge amount of accumulated experimental data; physics has a deep understanding on diverse phenomena and a vast knowledge of natural laws; profound theories explain many aspects of nature's behavior; consequently, there is enough data that solve the unification puzzle. The proposal contained in this study solves the problem by following a global logic analysis; firstly by investigating the most basic premises that may explain interaction among material objects; thereafter, it demonstrates the fact that we reside in a multidimensional time environment and that every observed object is made up with time structures, so constituting the organized universe we perceive. It explains most fundamental natural phenomena from a novel perspective, analyzing also some inaccurate considerations proposed by current theories.

This introductory chapter may appear more of a philosophical essay than a physics study, but it was oriented aiming for exhibiting the firm logic foundation of the TH. This proposition unites not only current incompatible theories in physics but also creates firm ties between natural sciences and basic philosophical principles; in this way it reestablishes the trust that every scientist must have on logic; therefore, by following this protocol, the TH discourages current tendency ignoring paradoxes and bizarre predictions by trusting almost dogmatically in mathematical results, no matter how absurd the predictions may be. This normally is consequence of exploring the limits of incomplete or imprecise mathematical relations describing natural phenomena.

## Current status in physics

The universe, as our senses perceive it, has been always the starting point for the study of natural phenomena; historically, all initial efforts were devoted finding the laws governing the interaction between clusters of physical entities such as "wind energy", "water waves", "light speed", "gravity", "radiation", etceteras; observing how they evolve by changing their form, their properties and their position in a three dimensional space. Initially researchers were no interested, and technology was not available, in order to investigate ultimate causes; in other words, attention has been focused on laws describing the causal relationships of macroscopic phenomena involving millions of particles. It is also important understanding that our language semantics reflects this perspective; our idiom is plagued with concepts and definitions that are consequence of the subjectivity under which we describe abstract entities as: light, gravity, fields, voltage, current, induction, space, magnetism, etc. Research efforts in physics until the Twenty Century have been focused into the practical target of proposing laws and equations relating causes and effects in the macro universe; in order to facilitate this analysis, mathematical tools are profusely used.

One must take into account that the validity of the results found by applying mathematical procedures and theorems is always limited by the implicit incompleteness of original statements or formulas describing it, which always have some grade of imprecision; neglecting these limits may derive in illogical proposals as: a black object whose gravity attracts everything, even the signals it sends; or remote particles that communicate instantaneously, etc.

Additionally, as explained before, some of the entities under investigation represent abstract objects, which are just a mental image derived from our sensorial perception; thereby, confusing substances with phenomena, or considering virtual entities as real, this approach may derive in dogmas, as: wave-matter duality of particles; electromagnetic waves supposedly constituting a substance that propagates through space at the speed of light; positioning objects with certainty in an inexistent flat space; gravity pulling everything, even affecting the light that travels at the same speed.

It is undeniable that there are some subjects from current theories in physics that urgently demand pending clarification. The paradoxical, though fruitful, conjecture about wave-matter duality deserves a deeper rational explanation, much more insightful than the feeble subjective interpretation on a supposed experimental observation currently supporting it; the uncertainty principle, undoubtedly related to imprecise questioning, also deserves a more profound justification; the successful proposition from Schrodinger's investigating atomic particles as standing waves also deserves a better explanation than its simple forecasting ability; the time paradoxes of the Theory of Relativity also merit elucidation.

Another factor influencing current theories arises from the common practice of accepting as real the results obtained by using mathematical tools developed to complement human logic processing limitations. The mathematics provides tools comparable to the assistance provided by a wrench in removing a nut. Even the more enlightened intelligence may find difficult reaching right conclusions when analyzing complicated syllogisms under complicated logic premises, or when existing nonlinear relation between them; in these cases is useful in assisting our innate intellectual limits by using mathematical theorems, logical operators and other procedural tools that will help us finding the consequences contained within the relations being analyzed; obviously the results will always be a "truth" but it may not always describe a "real" physical entity; this apparent contradiction is not due to mathematics limitations, but it results on the possible omissions or imprecise terms of the original statements, resulting in logical but unreal results. We must always accept that absurd is unacceptable when describing nature.

The cause of current situation may be due to the vast advances in theoretical mathematics at the beginning of the twentieth century, supported by the work of geniuses who discovered important relations, dazzled scientific community with their findings; these facts influenced current trend giving a supreme relevance in using mathematics as a prospective tool in nature's research. This attitude may also have been boosted by today's massive professionalization of science, forcing modern societies to provide planned education based in the pragmatic objective preparing capable professionals meeting the basic requirements from the labor market in applied and experimental physics; this approach causes a dogmatic attitude resulting from a paradigmatic education, additionally fueled by a highly regulated professional practice. This approach may result in a tendency to disqualify dissidence from accepted paradigms, and what is worse generates an indolent attitude when evaluating paradoxical results.

The contradictory situation is that there is a general consensus in the scientific community about the urgent need for innovative proposals, but when new ideas are proposed, they are required be respectful to current paradigms and should be presented in accordance with conventional format rules; this culture results in that new brilliant ideas may not be seriously considered by intelligent scholars, precisely because they are revolutionaries; additionally, sometimes excellent scientists keep new ideas for themselves because it may result too risky for their professional standing if they accept them publicly.

The Theory of Hylomorphism proposal is not orthodox, and emphasizes on imagination supported by formal logic, without resorting to mathematical manipulations unless necessary; most of its formulas are framed within an intuitive two-dimensional environment.

## Substance, essence, and contingence

Whereas the situation of incompatibility between current physical theories owes much to interpretations that sometimes confound cause with effect; or occasionally do not establish a clear relation between a natural substance and a contingent phenomenon, due to the understandable but inaccurate practice of conceding the category of fundamental natural parameter to what human sensorial perception dictates; "I observe it, therefore it is a basic natural variable"; this is the case in the concept of "light."

This introductory section aims for establishing a clear difference between concepts defining physical parameters, constants and variables that describe natural entities or phenomena, by correlating them properly by the causal relation between substantial and contingent entities; for instance, one cannot include contingent variables when describing the substantial entity that constitutes them; consequently, you can not include force, length, mass, energy or velocity in a formula describing particles, because is the particle's essence what defines these parameters.

It is also important for clarifying that one may risk losing objectivity when <u>prospecting natural entities</u> by using mathematical operators applied to incomplete formulas; therefore it is unwise to base scientific research in the indiscriminate use of analytical tools. It is important to clearly distinguish the essential from the contingent, as in the case of the space-time continuum; or the form from the substance, as in electric charges; or the cause from the effect as in the gravitational red shift. It is also imperative be always aware about the implicit conceptual fragility of composite abstract entities as: "space" or "field", because is incorrect considering them as physical stable entities; they all are just contingent objects generated by mental abstractions describing a contingent phenomenon; as in the case of "light."

The objects do not define the attributes or properties from their substance; quite the opposite, are the objects the ones that acquire the attributes given by their essence, without this key component any entity would cease its existence, because substance is the raw material constituting any physical entity, and it is not a mere attribute or property of the entity; quite the contrary, the form that acquires substance is what defines the individual characteristics of the composed entity. The relation between the perception of a phenomenon and the essence of the object involved in it is similar to the relation between symptoms and the disease; of course that the mission of medical science is for finding the causes and cure the disease, not only alleviate the symptoms; in the same way, the ultimate target of physical sciences is know the intimate interaction of substances constituting natural entities, not only find the relation between observable effects of groups behavior.

The TH denominates "Hyle" to the universal natural substance that constitutes photons, electrons, protons, neutrons, and other particles, as defined by its geometric characteristics. Interaction between single particles is the ultimate cause for every observed phenomenon involving composite natural objects, or subjectively defined abstract entities. Considering that all particles are independent individual entities, communication is a mandatory demand to structure a causally related material universe.

The simplicity of the Theory of Hylomorphism proposal is overwhelming because it reveals the infinite wisdom underlying the creation of such a complex universe assembled on the sole foundation of pure logic, guided by few simple laws governing everything. Just mention a case; the Theory of Hylomorphism demonstrates that the basic property of matter that defines all fundamental physics variables is elasticity and that the universal property of material particles is the constancy of their angular momentum.

When studying nature one should conceptually distinguish between the various natural entities it contains: individual particles as an electron or photon; complex entities as atoms or molecules; contingent entities as space, fields or waves. By following these principles, the Theory of Hylomorphism defines a primary substance, time; secondary variables as length, speed, energy, etc; abstract entities as space, fields, and others; all of them framed in a spatial environment generated by the presence of matter. The interaction between large groups of individual entities, which originate most of observed natural phenomena as: absorption, shock, or rotation; are just formal, not substantial changes.

## Reality, perception, and language

Perception does not always reflect reality; our brain builds abstract associative entities related to our sensorial perceptions by integrating causal phenomena; this is not unique to physics, it happens in any science: sociology proposes "middle-class"; oceanography "ocean waves"; physics "light"; medicine "symptoms"; all of these are abstract concepts could be described mathematically writing equations that relate causes with effects, but none of these equations will explain the substantial interaction originating the phenomena. This is why science objectivity demands clearly distinguishing between reality and perception, and the ultimate causes from their effects; substances from contingent entities; and individual interaction from averaged behavior.

One astronomer who observes stars and galaxies has a better perception of "reality" than a person attending to a theater performance or to the vision of a physicist that observes moving objects in his lab. While the delay in the arrival of the performer's image to the public is imperceptible, the existence of a finite speed of light is so obvious for the astronomer that it is more convenient for him specifying distances by the time that the signal takes reaching earth; furthermore, his perception that "space" is a byproduct of the travel-time of the signal is even more evident because he certainly knows that everything he sees happened long time ago, it belongs to the past, and the observed object no longer exists as perceived; consequently, he can never be sure of knowing the current status of the object he observes; the meaning of concepts like: now, location, and speed, become intrinsically imprecise.

Language semantics replicate perceptions, not always realities; when reading a book we say that it is at 30 cm from our eyes; it will be incomprehensible say that it is at 3 nanoseconds-light away; we also say that the wind is blowing, meaning that the air molecules move at a given average speed.

Another overlooked fact for the theatrical viewer in the previous example, is the way he perceives the positions, and movements provided by the geometric information of his stereoscopic vision. On the contrary, the astronomer calculates the velocity of the stars by knowing in advance the <u>physical properties of the emitting</u> element; this information allows to him find the relative speed by the shift of spectral emission, based in the assumption that the physical properties of matter are independent of its position. The astronomer, who confuses the concepts of space and time, confirms a natural fact of capital importance: that space is the geometric representation of delays in the arrival of signals from different transmitters, based in the observed fact that the signal speed is constant. Under this perspective, distance turns out be a secondary parameter derived from the delay in the communication of objects far from the viewer; if we assign a time delay for signal emitted by every source we observe, we will not need use the term "space." Philosophically, the construction of a universe without "space" allows us to seriously consider the hypothesis of a virtual universe.

Emitter's information is always contained within individual photons; some objects emit so many photons that it is easy miss "reality" because perception oversimplifies the phenomenon of light describing it as a continuous "wave", when "the reality" is that we perceive the average contribution of a large quantity of components. This picture produces a mental image that generates the abstract concept of "light" that could be described with mathematical formulas, causing light appear as a substantial entity causally related with the properties of emitters and the effects on receivers. This approach does not necessarily means that the abstract concept of light represents a substantial natural entity; light really is just a natural phenomenon created by myriads of particles involved in its generation, their travel and absorption. Analyzing mathematically these types of phenomena, one may predict consequences, but not always discern primary causes. Working with averages simplifies mathematical manipulation but losing deepness and significance. Consequently, for analyzing the physical universe one must clearly distinguish the average behavior from the individual interaction of substantial entities; for instance, gravity ignores individual interaction between the particles generating it. One must be always very careful differentiate the substance from the appearance, and the causes from the effects.

In Newton's time there was a clear difference between Physics and Chemistry, establishing a marked difference between substantial changes and the formal evolution; chemistry was limited in dealing with molecular interaction that changes the substance of material objects, while physics was limited in studying dynamic changes in objects positions. Modern science emphasizes on fields, overlooking essential interactions between substances; for instance: ocean waves are the shape taken by a massive number of water molecules subject to the wind, the equation that describes mathematically this phenomenon may be used for calculating forces and speeds, but it will not find the intimate properties of water molecules; similarly, the conventional concept of gravity ignores individual interaction between the particles generating it.

In short, a physical entity cannot exist if it does not contain matter in permanent communication; from this perspective, distance is simply a way of describing the delay with which matter communicates; consequently, space exists as long as it contains signals. Under the

Theory of Hylomorphism viewpoint, space is contingent entity whose characteristics are established by the properties from the signals it contains. The concept of a geometric flat space, defined as an expanse with absolute emptiness, is just a facilitating tool, not being a subject of study in natural sciences; it is important pay attention to the crucial importance of this fact.

Does the future exist? We will surely agree that the answer is "yes" when referring to <u>the concept of "future"</u>; but it is a name describing a potential entity; it must be in the "present" in order to "exist." Similarly we may question; does space exist? Theory of Hylomorphism states that space is not an entity with independent existence; it is just a potential entity subject to the existence of signals within it; consequently, space is no more than a useful entelechy to ease human communication; although it is difficult communicate "space" concept, not referring to "distance", and vice versa.

## Distance

Distance defines the separation between two objects; it is determined by the number of metric units contained between them. But it is important realize that the metric unit is always a conventionally defined unit, so is a subjective parameter, not a physical entity.

On the other hand, time is measured by counting the unitary intervals passing while an observed phenomenon takes place, but the total will always be a provisional result until the phenomenon ends; in the same way, present is fading permanently, and becomes past. Similarly, in order to measure the distance from one object one must assume that the signal travels at a "constant speed" and then find the delay from the moment that a signal was send, until receiving the acknowledgment from the observed object; even worse, we must measure it permanently because we will never be certain that the observed object remains in the same position; it may change its properties or positions. We will not always be able to predict its status, not because causal laws do not exist, but because of our lack of capability of instant communication.

An ordered universe implies the universal ability of matter to send and receive signals; every particle, atom, molecule, or object within the physical universe must receive, absorb, decode, and send compatible information.

Additionally there is another important point being considered: an ordered universe linking distant objects by means of communication carriers implies its ability in modifying receiver's properties at absorption, which in turn demands its capability for interpreting the frequency coded information communicated by the received signal. Consequently, the receiver must share a common time reference with the emitter, or at least they must accede to the same time reference; since both objects are independent; therefore, it is reasonable assert that both must contain an internal clock running synchronously with a master time Consequently, the receiver must share a common time reference with the emitter, or at least they must accede to the same time reference;  since both objects are independent; therefore, it is reasonable assert that both must contain an internal clock running synchronously with a master time clock[2]. The identical spectral signature of elements all over the universe proves this assumption. Also Planck experiments and Einstein's equations have shown that photons are oscillating entities whose energy is proportional to their frequency; consequently it is crucial for any receiver should read precisely frequencies.

---

[2] When a signal leaves the emitter, it is carrying coded information about duration, frequency, polarization, etc in its time structure. To measure objectively all these parameters, the receiver needs to count on a precise clock using the same time unit reference than the distant receiver. Without fulfilling this requirement it is not possible to establish a useful communication of information.

## Signal speed and delay

The delay in receiving a signal is proportional to the "remoteness" in a continuous, homogeneous and isotropic media; so it can be translated to distance by adapting the concept to human language semantics; delay must be converted to remoteness by applying a constant conversion factor 'c'. The value of this parameter 'c' can be easily found in a lab, by measuring the delay in receiving a signal travelling a previously set "distance", similarly to the experiment performed by Michelson–Morley. The distance may be arbitrarily defined and measured in conventionally (subjectively) established units.

After knowing the value of 'c' we will be able to find distances indirectly, not having to use a calibrated measuring tape; by using the conversion factor 'c' we can easily convert travel times to distance units; by applying this perspective, 'c' is just a conversion factor. Consequently, the real meaning of the relation between distance and time as given by this parameter is not, as normally interpreted, a speed; but a universal ratio. This analysis confirms the validity of the Theory of Hylomorphism proposal considering that "distance" or "length", is not a primary variable, these denominations for fulfilling human language semantics; our minds generated as a useful entelechy for locating geometrically relative positions as perceived.

Therefore, not being space a natural entity, it cannot be used as a reference in locating events; it is just a mean facilitating our understanding of communication delays. Our mind accepts more easily the concept of "signal speed" because we perceive that signal "moves." On the contrary, the Theory of Hylomorphism proposes that signals main objective is to "communicate information." It is understandable that this confusion arises because both: "speed" and 'c', are a "ratio of distance and time."

As explained, *the Theory of Hylomorphism establishes that 'c' is a "transformation factor" used for transforming delay times to* _arbitrarily set distance units_. The transformation from time to distance will be a common practice used in this text for fulfilling the human language semantics, additionally makes it compatible with the conventional concepts in classical physics.

Considering that delays are equivalent to the travel-time from photons going from the emitter to the receiver, and that this travel-time is independent of their relative direction, the transformation factor 'c' must be a universal constant independent of emitters position, establishing in this way a "space" that is "homogeneous, continuous and isotropic", whose "metrix" (distance ratio with a flat space) may change as set by each particle individual travel-time.

It is worth explaining that, while basic geometric optics establishes light paths considering that 'c' "change" within a transparent media, the Theory of Hylomorphism explains that this description corresponds to subjective perception from a distant observer referencing distances to a flat space, because really there is a denser "metrix" within the internal "space" of the transparent material.

## Communication and interaction

The necessary condition for the existence of a universe containing diverse individual objects is based in the ability of influencing each other, not only by communicating their location, status and properties, but sending signals capable to modify the internal properties of receivers in accordance with the received message. No active interaction denotes that there is not presence, since any undetectable object is unable influencing or being influenced by others; so it will not "exist" in the material universe. Consequently, all natural phenomena imply interaction, reacting to distant stimulus in accordance to information sent.

To elucidate this concept, let's imagine observing a cube containing two objects "A" and "B" in opposite positions. If no one of them emits signals, no one will perceive the other nor be influenced by it; consequently they will not be part of the same "universe." Nature is composed by individual entities assembled as an ordered conglomerate related by causal laws that rule their mutual influence. The spatial order implicitly requires exerting less influence on distant objects than to closer ones; this is automatically established by the geometric topology of central emission in a three dimensional space, by decreasing quantity of signal carriers per unit of area as they travel away from the source.

Referring to above example; when "A" emits signals always in the wrong direction, they will never be received by "B", the situation will remain the same, and they will not communicate, nor reside in the same universe. Considering that signal carriers are photons emitted by individual particles, in order to consolidate a material universe there must be objects which include a large conglomerate of particles emitting signals permanently and in all directions. Moreover, sending signals that will never be received does not make sense; at least one of the signals must find a receiver validating its existence, and the interaction must be permanent for supporting the fragile existence of communication, interaction, and "space." It makes no sense send something that will never be received; the notion of distance and direction fades away.

For an emitter which does not have the intrinsic capability in producing recognizable signals, and does not know if a given signal will ever arrive to a receiver, because it does not even know if there is an object "B" or where is it located, and will never touch "B"; then, the only way influencing possible distant receivers is that the signal carriers provide some other adequate information elucidating "remoteness"; as explained later, gravity provides this "remoteness" information.

Another noticeable aspect is that our universe contains diverse objects occupying different locations; to establish separation is imperative that the signals spend some time travelling between emitters and receivers. Instantaneous communication would place everything in a point, destroying the possibility of building diversity. Additionally, the communication delay must be proportional to the remoteness.

A controlled experiment, will demand placing a mirror at "B" reflecting a recognizable signal emitted by "A", in the same way that Apollo astronauts left a reflector on the moon's surface; then, the distance "A-B" could be calculated by measuring the round trip travel-time of the signal, and multiply it by 'c', but nature does not work like that; even when individual particles receive permanently signals from each other, these signals are not recognizable, and the receiver will never be informed about the total travelled time. Then, it is reasonable assuming that the geometry of the particles allows them distinguish directional information (they must not be spheres), that signals include remoteness information (gravity must exist), and that the interaction mechanics is not based on individual signal recognition (signals depend on the emitters mass, not from composition).

## A single substance for everything

Simplicity is the master rule of nature, under this premise it would be unwise explain universal diversity with a variety of ingredients as: particles, matter, force carriers, gravity, electric waves, magnetism, time, fields, space, etc.; suggesting that there must be a common raw material which constitutes all natural objects.

Additionally, as it has been proven above, there must be a necessary interaction between objects residing in the same universe implies that all particles are made with the same substance because if so, all them will have the ability to emit, absorb and interpret information carried by signals; this assertion suggests that the difference between basic particles is merely formal (size, structure, form). The TH denominates this substance as "Hyle" (raw material in ancient Greek) as a deserved tribute to the genius of Aristotle, who twenty six centuries ago envisioned nature's Hylomorphism.

Considering the current tendencies in physics, it is highly predictable that many scholars will tend disregarding present proposal; as simplistic and with no scientific value. This may be comparable to the attitude from a nineteen century biologist judging as simplistic and irresponsible suggesting that all cells in a given organism are formed in accordance to the structure of a particular acid, the DNA.

Trying to explain that particles ability to send and to receive signals by assuming that they have a signal's deposit, and that they will change their properties depending on the "amount" of that substance stored in it, is not realistic; since if they ever run out of the signal substance they lose their ability to emit signals, demanding to be replenished before recovering that property; it is also unlikely that their properties will change with the amount of stored "communicating substance". This analysis suggests that the simplest solution is for signals to be structured with the same substance than the one constituting the particles; their Hyle "amount" increases, when they receive (absorb) a signal and decreases when they emit it.

An additional conclusion will be that the particle's substance builds elastic geometrical structures, whose individual dimension changes when the particle gets perturbed by some external cause, and that it will try to recover its shape and size, to remain being the same.

In accordance to above description, the Theory of Hylomorphism postulates that it takes only one substance (Hyle) for assembling a unique structure, which must be governed by the same simple rules and may acquire a range of shapes and properties depending on the particular structure shape, a photon, or matter; which in turn shape space, light, atoms, molecules, stars, and planets. Furthermore, these laws were established so wisely that matter evolves to life forms and eventually consciousness.

In this way the TH proposes a unifying theory; the unifying quest in physics started when Newton established the relation between mass and weight, followed by Einstein who gave a step forward establishing the mass-energy equivalence. For the TH, mass, energy, force, momentum, etc are terms always referring to some formal property of Hyle.

## Hyle, an oscillating substance

Transmission of complex information as: remoteness, charge, polarity, and direction, cannot be coded by using only one parameter; as the amplitude which ultimately is only one variable; not even two independent variables will suffice. Nature offers five independent variables (four time dimensions and amplitude), that signal carriers use to encode the information they will eventually communicate to a receiver; three of them compose a vector of given amplitude pointing to some direction, and position it in some spatial location at a given time; those parameters may be ruled by a law. Consequently, it is reasonable assuming that the substance (Hyle) constituting the signal carrier (photon) must contain tridimensional vectors changing direction while moving away from the emitter.

Under this model, rotation provides absolute time information, while spatial information uses the vector's plane of rotation relative to the displacement direction. But a sole vector is just a point positioned in the 3D travel-time space; it does not constitute an object with spatial dimensions. An spatial entity must have at least one dimension, has to have a length, which means that it must be build by a sequence of vectors sequentially positioned on the displacement straight line [3] set

---

[3] Is a straight line because as it increases its distance to the emitter at a maximum rate; this is the definition of "straight line."

by its trajectory. The simplest geometric description of this entity resembles a helical structure similar to a screw.

The Theory of Hylomorphism proposes a binary vector whose amplitude is one within the particle limits; a vector with magnitude other than one or zero is useless, as will be shown.

Under the TH perspective, the distance to the emitter increases as the travel time of the signal carrier increases on each revolution (set by its length 'T'), similar to the way a screw penetrates with every turn. Considering that the total time traveled by the signal is proportional to the number of revolutions, and that the total travelled distance (travel time) in that time period will be given by the summation of advanced "length" 'T' per revolution, it is easy to intuitively understand how distance unit (metrix) may change, "bending" space as a function of 'T'.

## Hyle structures

Any oscillation in a three dimensional spatial environment is related to a vector which changes direction continuously, so the projection of its amplitude on each of the three dimensions varies accordingly. This vector moves away from the emitter as time passes.

The TH proposes that Hyle, the substance constituting all particles, *"is a structure which rotates permanently while travels away from its origin"*, always increasing its travel-time by one length per each complete turn; the period (frequency) of the signal and its length is defined by the emitter. Therefore, the substance building particles could be described as: a <u>continuous binary vector function</u>, assembled as a moving structure with a travel time "length" 'T' and a rotation period 'T*'; this structure will always travel away in the same direction, advancing a travel-time 'T' on each revolution.

Consequently any particle is a one dimension entity with a given length that travels in one direction as it rotates. Therefore, the components of any Hyle structure are:

- **Time**; it is the ultimate substance from which everything in the universe is made; it has two components: a) absolute time which sets its rotation period, and b) the travel-time which establishes its length.

- **The length of the structure**; defined by the travel-time increase per revolution. (Similar to the pitch of a screw)

- **A rotating continuous sequence of binary vectors**; with amplitude equal to "one"; defining the beginning and the end of the structure.

Under the TH perspective, time is the ultimate raw material that constitutes everything, and the everlasting substance of a causal universe; it rules everything that exists. The universal time runs progressively and constantly; absolute time advances independently of the position, shape, change and velocity of any object, while the communication delay, defined by the travel-time, depends on the relative isolation of the interacting particles.

The other ingredient is ontological; it can be described as a vector function whose amplitude is binary; taking the value of "1" or "0" (exists or does not exist in a given point in space and time) while it travels away from the emitter rotating permanently at a constant angular velocity. Its rotation constitutes the individual universal time clock within every particle. Considering that this vector function define a structure that rotates as a whole, increasing permanently the travel-time from the emitter, it institutes the spatial unit of "length" in a particular point at the 3D travel-time space.

## The origin of elasticity

As it will be shown later, an entity as described above could be subject to external influences that may modify the relation [4] between 'T' and 'T*'; considering that every segment in the vector function constituting the particle must comply with the law ruling that "it must separate from its emitter a travel-time equal to absolute time elapsed, and the front part of a particular particle may move less because it is pulled by the tail, this difference will generate a "stress" due to the structure's tendency to recover its natural "length" by reaching a symmetry between the length 'T' and the period 'T*'. This explanation helps in elucidating the origin of "force" as the fundamental parameter in physics; this perspective explains how an internal distortion on the ratio between the two time dimensions is closely related to the concept of "force." The same concept applies to any complex material particle structure which will always react as a spring.

## The concept of velocity

The concept of velocity in accordance to Newtonian perspective of nature is the ratio between a distance that the observer measures with a measuring tape, and the time that takes a moving object going from one point to the other, as given by the observer's clock. Obviously the observer that monitors the motion is always far from the moving object and he cannot always measure the distance directly, so the interval he records by using his own absolute-time clock is distorted by the delay of signals departing from the first position and the one coming from the second; consequently the term "velocity" in physics is always relative and referred to the observer's subjective perception on a distance travelled by the moving object "measured" from far apart and time intervals computed locally including signals travel time; so the definition of velocity is prone

---

[4] The travel time dimension of the particle is a geometric property which depends on the space "curvature" where it resides.

generating ambiguities when not precisely specified [5]. The TR explains this concept in full detail.

## Continuous vs discrete signals

As already explained, it is reasonable assuming that any particle must be able to: absorb/store/send them. It could do it by having a suitable reservoir for every type of signal (one for storing mass/energy, another for electromagnetic data, and another for gravitational information, etc.); it is much more reasonable assuming nature solving the problem by using the same material for constituting the signals, the material particles, and the containers. Consequently, the storage of signals should produce a formal change on the receiver, producing an instability which may result in the emission of the excess of substance. The two important consequences of this assertion are that:

a) Signals must be discrete entities, otherwise the supply will be promptly exhausted by a continuous emission; additionally, as the signals are made with the same substance than the emitting particles, the common raw material may exhaust and the particle will disappear shortly; even worse, when the particle spends its own substance, it will gradually change its identity until finally disappearing; under this scheme, the emitter will send information on itself that varies as time passes; so it is rational concluding that emitting particles must return to its initial conditions after every emission; consequently, the emitted signal will always be a discrete packet of surplus substance.

---

[5] As the moving object changes positions in a time period, the observer will always perceive that objects moving away increase artificially their "distance." Considering absolute time measurements, recorded time interval will be distorted because two orthogonal time dimensions are added; the absolute time of the moving clock and the increase in travel time of the signal. The Theory of Hylomorphism demonstrates that Lorentz transformation does not use absolute-time to calculate relative velocity.

b) A particle must send only one signal in a given direction at a time, so complying in informing always about its stable status characteristics. Consequently, it is necessary count on massive bodies which send permanently large quantity of signals in all directions informing about the average status of the conglomerate. The temperature of the emitter is proportional to the internal interchange of signals and consequently also to the flux of signals leaking out per unit of time; this fact also explains the third principle of thermodynamics by showing that the internal interchange of signals decreases permanently because a portion of substance travels away forever, as it will never return to the emitter.

The above facts were experimentally confirmed as stated by the Einstein-Planck relation, proving that signals are energy packets known as photons. Surprisingly, this cautious assertion was carefully phrased as "energy packets" so avoiding confrontation with the universally accepted concept, that the electromagnetic essence of light is a continuous wave, which was the dominating dogma at that time. At the end of this book, Annex I analyzes this point in full detail.

Therefore we may conclude the discrete quality of photons. Light intensity is given by the quantity of photons arriving to a surface, not by amplitude of the individual photons; while energy is given by the light wavelength.

## The need for a three dimension space

Any natural entity has to have a size that occupies part of a dimension at a given position; additionally could change in form or substance, and could form complex objects. A one dimension universe could only contain segments and could only communicate information sequentially; a 2D universe may have bi-dimensional objects that will vary in form and area but cannot build complex shapes in diverse spatial location, able to interact accordingly; a three dimensional space offers the minimum adequate room for building complex individual objects located in diverse places interacting permanently. The fourth natural dimension is absolute time, which provides the ability for instituting a causally induced change.

The perspective proposed by the TR instituting a space-time continuum as a consequence of considering the "speed of light" as the fundamental natural parameter and its mathematical version, offers "truthful" results about space by predicting spatial realities experimentally proven, as the space curvature; but it also presents some misleading concepts about relativity of time and its relation with space, this incoherence is originated by considering "space" as a substantial natural entity, so blending properties of time dimensions.

By subordinating to the concept that 'c' is a substantial nature's parameter, experimentally proven by the light speed constancy, the TR bases its development on the mutual dependence between distance and time, effectively mixing the properties of absolute time with the ones of travel time. Under the Theory of Hylomorphism perspective, the travelling photons establish the distance unit (relative to a flat space); known along this text as the "metrix" of a curved space. The Theory of Hylomorphism establishes that light speed constancy is a consequence of the subjectively defined distance units, and establishes that 'c' is a conversion factor which transforms travel-time units to distance units[6]; remember that travel time is set by the length of traveling photons. Under this perspective, "space" is a contingent entity, "light" an abstract concept and "light speed" 'c' results an improper way to perceive the signal travel time.

---

6  Unavoidably we have to use some imprecise denominations, as "space", "light speed" and other conventional terms, because semantics of human language obliges us to do so.

## Frames of reference

The Theory of Hylomorphism establishes that the only absolute point of reference is always the observer, because it is always the observer who receives information about the neighboring universe by absorbing signals emitted by distant objects. Complex natural entities are also composed by individual particles interchanging signals that inform about their mutual relative position and their internal changes.

Remote particles never inform directly, nor instantaneously about their distance; the signals do not have an internal travel time counter neither stores the relative angular position of the emitter; the distance can only be perceived indirectly by arranging sensors geometrically. Considering those facts, to set a reference system at some point in space other than the receiver would mean assuming future position/velocity of the entire environment[7]; any intermediate point between emitter and receiver does not have a physical meaning, since not even the emitter knows if the signals he sent already arrived, or will ever arrive, to a receiver. Moreover, the receiver is also always stationary; it cannot change relative position within a "flat space"; the classical mechanics approach referencing to diverse coordinate systems erects feeble references. It becomes absurd refer to a position when it is impossible to "mark" the previous location because the rest of the universe is changing around it; permanently redefining the space properties at all points. Curiously this statement returns physics fundamentals to pre-Galilean assertions. Large masses emitting signals continuously establish a surrounding environment that is a resemblance of "space positions", but always dependent of the emission's stability and the relative influence from other neighboring objects that may change relative positions and internal properties.

---

[7] This stable condition is only given in nature around an emitter sending signals continuously.

Consequently it is not an advisable practice studying physical phenomena by subjectively setting various "frames of reference in a flat space"; one should always consider that by doing so, one is implicitly assuming the stability of the surrounding environment. The truth is that the future status in surrounding environment will always be uncertain; although, most of gravitational environments are stable enough for applying concepts established by the General Theory of Relativity and in the Newtonian mechanics.

## **Matter and signal interaction**

Under the current classic perception of nature, material objects contain large amounts of spherical molecules building solid objects with a defined shape, or dispersed in a volume as liquid or gas; signals are electromagnetic waves which illuminate those objects. From the Theory of Hylomorphism perspective both entities, particles and signals, are constituted by the same substance and both will have only one spatial dimension (length); their difference is that material particles are rings, whose Hyle keeps circling around a circular path, while Hyle within photons travels permanently away from its source. In the case of particles, the closed path defines its own circular space where the straight line is a circle; in the case of photons, Hyle defines a straight line as its path; in both cases the Hyle travels permanently on a straight path but within a different spatial geometry. Therefore, material particles may <u>reside</u> permanently in external "space" location, while photons travel within it.

When a photon finds a particle, it penetrates a ring "own-space", so it is absorbed while keeping its straight path but it becomes immersed in a circular space circling the ring together with the Hyle of the particle, so it will appear as winding into the ring. The opposite happens when a signal departs from an emitter.

Like in any encryption systems, the signal carrier communicates time-coded information to particles; therefore, the carrier must keep an accurate timing with its own clock in order to be able delivering useful information to the receiving particle which also must have its own universal time clock. This analysis confirms that without a common time reference, it will be impossible in exchanging useful information; decoding time-coded signals always rely on appropriate synchronization.

The signal carries information in two ways: a) Dimensional information: changing its own length (travel-time) while moving away from the emitter; and b) Structural information: the direction of rotation and its rotating plane will define the influence on the receiver; as said before, reading information demands both particles have a precise universal time clock. It is important to remember that: a) photons are individual "signal carriers", so the receiving process is always performed locally by <u>one photon with one particle</u>; the interaction lasts while all the photon is absorbed; and b) Photons do not store elapsed travel-time information, so they communicate delay time between receiver and emitter by giving gravitational information; it will be impossible build a universe containing diverse objects, without a gravitational field.

Another important fact is that in a three dimensional environment, the probability for a single signal emerging from an individual emitter arriving to one specific receiver is very low; consequently, nature demands massive conglomerates of emitters, sending constantly a large quantity of signals in all directions; nevertheless, the interaction keeps being local between two individual entities; a photon and a particle.

In accordance with TH, the fundamental natural law is that angular momentum from matter rings is a universal constant; so, massive particles are small, and those with less mass are larger. Other internal properties from particles are defined by the relative direction of the rotating plane from their Hyle.

## Natural phenomena seen under the new perspective

The Theory of Hylomorphism analyzes some basic subjects from current theories in physics[8]. Some unconventional vision and conclusions derived from the proposed perspective are:

- Nature resides in a four-dimensional time environment; that is why the proposal is entitled as "The Physics of Multidimensional Time."
- Structured time is the only substance constituting material particles and signal carriers. This substance is known as "Hyle" as a deserved tribute to the genius of Aristotle.
- Space is a contingent entity, resulting from a mental abstraction. It is contingent because it ceases to exist when there are no signs crossing it; consequently, the distance is a concept derived from the communication delay between particles, therefore it cannot be part of the substance constituting physical entities (the effect may not be the substance of its cause).
- Particles are individual entities with primary properties (form and frequency) that differentiate each one from the others, they also have secondary attributes as: elasticity, Spin, mass, energy, etc. All material particles are circular loops of Hyle whose electric charge is defined by the direction of the Hyle rotating plane.
- Elasticity of particles originates the concept of force as the internal spatial property restoring dimensional travel-time symmetry; gravity is a consequence of this tendency when photons emitted by a massive object generate a velocity

---

[8] A brief comparative analysis on the similarities and differences of the Theory of Hylomorphism with each one of the main theories is discussed in Annex 1.

field (bending the space) as they decompress while travelling.

- Structured groups of particles organize producing complex structures such as atoms, molecules and bodies. Atoms are formed by electron rings centered on the nucleus, which is a basket, woven with protons and neutrons.

- There may be groups of particles forming abstract contingent objects such as: space, electric field, the gravitational field, radio waves, spectrum, induction, etc. The Theory of Hylomorphism shows that some of these entities are abstract entities, such as: magnetism, electromagnetic waves, etc.

- The erroneous concept assuming that "magnetism" is a natural substance was engendered by wrongly assigning a quality of physical entity to an intermediate vector variable used for simplifying the mathematical and geometrical analysis of the phenomenon of forces applied to a particle moving in any "velocity field." Magnetism is not a natural substance.

- "Fields" are abstract denominations resulting of mathematical formulations from an abstract concept describing the geometry of <u>the potential effect</u> exerted on a particle by a distributed group of distant particles.

- Electromagnetic waves are the geometric description of the field produced by synchronous oscillating sources orderly distributed in space. Maxwell equations derive from mathematical theorems (Gauss T, T Stokes and equipotential distribution); therefore, the emitters may be material particles, electrical charges, or some other type of radial force; they may also describe an elastic process as a group of swirls; the equations are exactly the same in all cases, consequently an electromagnetic wave is not a physical substance; the same wave equation may describe the "magneto-gravity" generated on a self propelled spaceship, or the forces acting on a cork floating close to a drain.

- The universe was born when a powerful light was emitted at the beginning of time, as a massive emission of Hyle, setting the form of myriads of high energy photons; some of them curled on themselves forming neutrons which eventually broke as protons, electrons, and photons; all these particles structure the atoms, molecules, material objects, gravity, space and all the universe as we perceive it today.

This introductory chapter, as pointed previously, may appear being more of a philosophical essay than a physics proposal, but it is oriented to show the solid logic foundation of Theory of Hylomorphism proposal, unifying not only current incompatible theories in physics but also creating firm ties between natural sciences and philosophical principles; pointing to an alternative route to some current tendencies that ignore paradoxes and bizarre predictions by trusting dogmatically in mathematical results.

## The natural system of units

Finally, everything in nature is made only with one substance, the "Hyle", which resides in a four dimensions time environment; therefore, the magnitude of all the variables, parameters, and conversion constants must use the same fundamental universal unit, "the second." Distance units are TT [seconds], velocity is a non dimensional factor; the gravitational constant "G" is in $[sec]^3$; mass and energy in $(1 / [sec]^2)$.

The conventional "systems of units" size variables profusely referring to arbitrarily defined units, so does not give a profound insight on the intimate essence of its natural significance. On the contrary, the natural system of units provides time powers as unitary magnitudes;

this may enrich the conceptual relation between supposedly independent parameters sharing the same unit.

This page is intentionally left in blank

# Chapter II

## The four time dimensions

### I.- Space

The introductory chapter explained the indispensable need for communication in nature; arriving to the conclusion that the ability of matter in receiving signals must be universal; every particle, atom, molecule, or material object in the physical universe must be able to: receive, absorb, decode, and send compatible information. This chapter investigates the substantial characteristics of the "remoteness", as a variable that specifies separation between diverse material objects, and how it relates to time; explaining also how the concept of time relativity proposed by the Theory of Relativity is the consequence of blending the properties of two time dimensions.

### Reality

Our senses detect natural signals communicating to our brain about the conditions of surrounding environmental conditions allowing us perceiving sounds, force, chemical composition, heat, and light. Our senses are specialized transducers that interact locally and send all perceived information being processed under logic premises, producing virtual entities that correspond to abstract object or phenomena as: wind, wet, color, and other names we give to them subjectively; not always being aware if they are objects, substances or phenomena in permanent evolution, as temporary entities whose existence is always transitory.

We, instinctively locate an object employing a mental Cartesian coordinate system centered in our eye and establishing a point in front of us at a known distance as a reference for lateral positions, elevation and remoteness. The first two coordinates (lateral and elevation) are easily perceived and can be found by a simple geometrical calculation when we know our distance from the reference point; but remoteness is perceived only by the size of the object, their distance is related to our perception if it is small or large; this fact demands knowing previously its dimensions in order to find mathematically how remote it is; in all three cases we must know previously the size of the observed object and the distance to the reference point, and we must apply mathematical procedures in order to find positions; obviously, nature does not work with such parametric information.

In a lab experiment, a researcher may always mark previously reference position by using a calibrated measuring tape, so he can perceive almost instantaneously the moment that a moving object reaches every point; but when the observed object is not within reach, like the moon, he cannot measure and mark directly distances; the only way in knowing the distance to an object on the moon, is sending a recognizable signal that will bounce back allowing him find the total travel time of the signal, using this result to calculate the distance, assuming that he knows previously the value of the signal speed, ad that is a constant; if he uses a luminous signal, he must know the light speed and make sure that it is constant.

Due to this reason, the astronomer who observes stars and galaxies has a better perception of "reality" than the physicist experimenting with nearby objects moving within his laboratory, because the delay in receiving their signals inside the lab is so small that may be neglected. On the contrary, an astronomer knows that a finite speed of light is an ontological requirement that nature establishes setting a delay proportional to the remoteness in all directions; is that so, that for him it is more convenient and coherent for representing distances by the travel-time of the signals. Effectively, he considers "distance" as a byproduct of travel-time; fact that additionally indicates him that everything he perceives happened a long time ago, belonging to the past, so it no longer exists as perceived; consequently, he fully understand that he will never be certain knowing the current status of what he observes; under this perspective, the meaning of: now, there, and velocity, becomes fragile, relative and imprecise.

A lab researcher perceives positions and motion as <u>geometric information</u> used to find distances and velocities by applying mathematic theorems and procedures; on the contrary, the astronomer <u>can use only physics</u> (natural laws) in finding object position, and its velocity; he does not have access to any useful geometric reference, so he calculates the speed of the galaxies by knowing in advance the physical properties of the emitting element and some natural laws as the constancy of light speed for transforming delays to distances, the spectrum of stars allows him to find relative speed by the shift of spectral emission (he assumes that matter properties are independent from their position and speed); the Hubble law allows establishing the relation among speed and distance. In this manner, an astronomer, by confounding the concepts of space and time confirms a natural fact of prime importance: distance is a 3D geometric representation of delay in the arrival of signals from different transmitters; under this perspective, distance turns out being a secondary parameter derived from the delay in the communication of objects far from the viewer; if we assign a signal delay to every source, we will not need to use the term "distance" and we can build a three dimensional environment of delays. Philosophically, the construction of a universe without a preexistent "space" allow considering the hypothesis of a virtual universe.

## Space, a contingent entity

Imagine a cube in complete darkness containing two objects 'A' and 'B' in their farthest extremes; both can only be seen by an outside observer; if neither of them emits signals, the other cannot detect it neither be disturbed by the first; the absence of interaction implies that none of them is subject to the influence of the other, so they will evolve independently, therefore is reasonable saying that they do not reside in the same universe.  If 'A' send signals that 'B' do not receive it, the situation will remain the same; even worse, does not make sense issue a signal that no one receives, since it would be very difficult explain how one may send something being sure that will never be received [9].  This example shows that sending something traveling a pre-existing "space" is an absurd when there is not at least one receiver; this argument intuitively reveals the fragility of the concept of "space", which vanishes when the volume does not contain entities in constant communication.

Describing space to a blind person one might say; "is the time that takes a signal finding a target", obviously assuming that nothing is in-between; this example shows clearly the subjectivity of the concept of "space", and the obvious incongruence on including it as a primary parameter in a physics formulation.

Once more we conclude unequivocally that for physics, space is a contingent entity, and a secondary parameter whose existence depends on the presence of signals communicating material particles. Facing such strong conclusion, the concept of "distance" deserves being analyzed since it is the parameter defining the reference scale for every spatial dimension.

Does the future exist? We will surely agree that, as a concept, the answer is yes, is a contingent entity that must exist at present time, to become "real"; similarly, does space really exists? Theory of

---

[9] (*) On the contrary, asserting that the signal leaves an emitter and increases its travel time, is logically acceptable.  When utilizing a spatial perspective it is absurd is to state that the emitter sends a signal because one  implies the existence of a destination, while focusing on the signal perspective, exploring does not need a receiver; only if there is one, the space concept could be applied; if not it remains being just a potential entity.

Hylomorphism states that space is not an entity with independent existence; it is just the consequence of the existence of signals passing through it; when it does not contain signals it is like future, just a contingent entity; consequently, we may conclude again that space is no more than a useful entelechy, not a substantial natural entity.

Another topic that should be remembered is that matter needs three dimensions in order to structure complex objects occupying diverse locations; as previously shown, this objective is not achievable with only one or two dimensions. Additionally, nature must have the tendency to structure individual clusters of matter, in order to have the capability of permanent communication, so constituting a causal universe ruled by laws.

**Distance and delay time**

Particles recognize each other's existence when both receive the signals from the other, but they cannot evaluate the distance A-B previously set on discretionary basis by the experimenter, because they do not have way of knowing the travel-time of the received signal, unless they could recognize somehow their own returning signal; let's suppose that this is the case. Once having signal travel-time, we can calculate distance A-B by applying the conversion factor 'c' that must be a constant related to the metric system of units used in the experiment, independent from the emitter properties (mass, internal characteristics, velocity, acceleration, composition, or any other); otherwise the universe would be chaotic, since the observer might absurdly receive first signals from remote masses located at privileged positions; or from massive emitters sending faster signals; or from approaching objects increasing the speed of their signals.

For those who do not trust in this pure logic analysis, the constancy of speed of signals is an experimentally proven fact; it is found by measuring the time it takes the signal for traveling a preset distance. The result of these experiments offers the equivalence between the "distance" measured using an arbitrarily set unit, and travel-time; it erroneously refers to the "speed of light" 'c', assigning the concept of velocity to a parameter that is just a conversion factor relating casually the same units; additionally, this perspective was influenced by the semantics of our language that forces us referring to "distance", when we should mean "remoteness."

For mnemonic reasons, the above description simplifies the explanation by assuming an abstract flat space, which does not exist in nature.   A flat space implies that it does not contain matter; consequently it does not exist.  But "space" will always be the "media" which our mind uses (entelechy) for establishing that there is a time interval separating the emitter from the receiver; The term "distance" is used in future explanations in this book, to avoid adding unnecessary complications when explaining concepts to the reader, who always must keep in mind the abstract meaning of the terms "distance" and "space."

Consequently, it may be concluded that physical phenomena includes interchanging information between matter conglomerates; producing a reaction to this stimulus on the receiver, always in proportion to the quantity of information carriers received; consequently, distant or dark objects exercise less influence than closer or brighter ones.

## II.- TIME

Time elapses continuously; we only know momentarily the universal hour (at present) as its total elapsed amount; which always disappeared in the past. We measure time by counting the number of arbitrarily defined unit intervals, related to the rate of change provided by nature.

## Absolute time

When observing natural material and immaterial universe, we find that it is unconditionally subject-to the principle of causality, all phenomena have a cause producing always the same effect, and what is more important the cause always precede the effect; therefore it is reasonable assuming that the main natural ingredient is the absolute time; this assumption will prove being nature main cornerstone, even at particle level. Lacking universal absolute-time will be impossible to differentiate cause from effect, and physics as a science could not exist; the universe will be a chaotic entity prone for collapse because of a missing universal reference. The principle of causality also establishes a very important rule for physics: communication cannot be instantaneous and therefore should always be a previous cause producing a subsequent effect. The objective of physics is precisely to find the rules which link causes and effects.

Time is not only the key ingredient for causality but it also provides the basis in structuring a diverse universe; without time, space cannot exist because time provides also the platform to establish a foundation to differentiate between phenomena happening far away and those occurring nearby; communication is faster and more frequently between closer objects. Moreover, time defines change; from the smallest particle to the largest galaxy everything evolves changing shape, position, and condition.

We live in a universe in which <u>nothing causally related happens simultaneously</u>. Simultaneity is always apparent; related to the particular perspective of the observer position which must be equidistant to both sources for detecting it. There may be some cases when an observer receives simultaneously signals from sources that are located farther one from the other; consequently, observed simultaneity is always a subjective appreciation.

Natural order speaks about the unconditional and universal adherence to the principle of causality; any phenomenon that appears being chaotic is a mere confirmation of our limitations or ignorance about in-deep knowledge on the laws relating its cause and effect. The concept of causality implies the need for absolute time to be: universal, sequential, continuous, homogeneous, and irreversible.

As it has been shown before, travel-time is an independent variable from absolute-time. The Theory of Relativity proposal, assumes time relativity by mixing both variables when considering space as a separate substantial entity, and light as basic parameter. In general, the misleading notion of only one time dimension may lead to "unreal" conclusions.

## The bi-dimensional quality of time

The time interval '$\tau$' that takes a signal travel from the emitter to the receiver, denominated as "travel-time" or TT in this text, <u>is independent</u> of the passage of an absolute-time 't' running autonomously; the signal travel-time only depends on the emitter-receiver relative remoteness. The concept of the independence of a signal's transportation time could be explained in a more conventional, but improper way by saying that '$\tau$' depends only from the distance ($\Delta s = c\,\tau$) between emitter and receiver, assuming that the signal travels at a constant speed.

The term "distance" does not refer to a predefined parameter; it acquires a meaning only when a receiver detects a signal, answers it, and receives the acknowledgement; but considering that nature does not behave in accordance to this protocol, the receiver may perceive similar signals arriving from the same source, but it will never know which one is the expected reply.

Considering the difference between "distance" and travel-time, it is fundamental considering the fact that nature is a universe of times which occupies four independent dimensions including three travel-time and one absolute-time coordinates. The vector position 'r' is defined by all four coordinates as $r(\tau_x; \tau_y; \tau_z, t)$ where '$\tau_i$' are the travel-time coordinates at the conventional directions (x, y, z), and 't' the absolute-time. We may then write $r(\tau_i, t)$ in order to explicitly show the important difference between the properties of the three travel-time coordinates '$\tau$' and the absolute-time 't'; so this proposal refers to the "two dimensions of time" or bi-dimensional quality of time, because considering that there are two conceptually separated time dimensions; which additionally, will suffice in the analysis of most natural phenomena.

Therefore, qualitatively the two dimensions of time are:

- **The "absolute-time" (AT)** 't' rules the rate of change for everything, so establishing a universal reference. The absolute-time interval is the universal reference giving the capability for synchronizing the interactions between every particle of the universe. The smallest time interval is set by the fastest natural oscillation (neutron?). The absolute-time is continuous and irreversible; its unit interval is the reference set by nature. The past is always increasingly positive and the future is a contingent variable that cannot be measured, can only be presumed.

- **The "Travel-time" (TT)** '$\tau$' is the elapsed interval from emission to reception. It is a different kind of variable; travel-time accumulates time intervals while the signal travels from the emitter to the receiver; it is a three-dimensional variable because the signal may arrive from three independent directions. For the most of this study, travel-time is considered a single dimension variable, taking into account that including all three dimensions may complicate unnecessarily the analysis without providing additional conceptual clarity. The travel-time of a signal is a discrete variable[10]; it can be positive or negative; it is homogeneous, and reversible.

This text refers to "space" in the conventional meaning, considering that it will be difficult communicating positions without using the terms "distance" and "space"; due to language semantics, but always one should remember that they are nonexistent natural entities. Similarly, it will be difficult mention the moment that takes place events without using the term "at that time"; meaning "the universal hour of the event." The irreversibility of absolute-time implicitly establishes the existence of an "universal-hour"; permanently increasing since the beginning of times, making it possible to mentally set a universal reference to communicate the time of occurrence of an event; otherwise it will be complicated talking by saying "as perceived by an observer capable in knowing instantaneously the time given by all clocks in the universe" or by saying "for an observer located equidistant from the emitter and the receiver, that can always receive simultaneously the actual time marked by their individual clocks." The universal-hour is an existent physical entity, but it is impossible to experimentally prove that its rate is a constant, because is the parameter mastering all natural sequence.

As everything in universe permanently changes place, the only functional coordinate reference, will always be the observer's position. As seen in FIGURE I, it is not possible to establish an invariant point on this two-dimensional coordinate system, because even the "travel-time axis" moves permanently, changing always the origin of the "present", continuously set by the current value of "absolute-hour."

---

10 The "space" travel time resolution depends from the signal "travel-time length." Its smaller fraction is given by the minimum <u>length of a non-destructive</u> high energy photon.

The three independent travel-time dimensions represent a subsidiary dimensional design used by nature for building complex spatial structures; the natural phenomenon is always a one-dimensional phenomenon referring to the movement of an individual signal.

The physicist must always be aware that observed phenomena are past event. An observer will never directly know the absolute-hour when a signal was sent; he only knows the absolute-hour when it was received.

## Time addition

When observing the change of position of an object travelling away from the observer located at the origin "o" who sends a signal from a new position (o + $\Delta$s) after a time interval '$\Delta$t' as measured by its own clock (absolute time); the communication delay between the two signals will increase by the magnitude of '$\Delta\tau$' [11]; so it will be perceived by the observer after a time interval [12] that is a vector '$\Delta\zeta$' resulting from the addition of two orthogonal vectors:

$$\Delta\zeta = \Delta t + \Delta\tau \qquad (2-1)$$

---

[11] (*) The same increase of delay will be observed for an object departing from a distant point "o" and travelling away to (o + $\Delta\tau$).

[12] This effect is masterly explored by the TR, but based in the wrong premise that the phenomenon takes place in a space-time environment considering only one time dimension; blending absolute time with travel time.

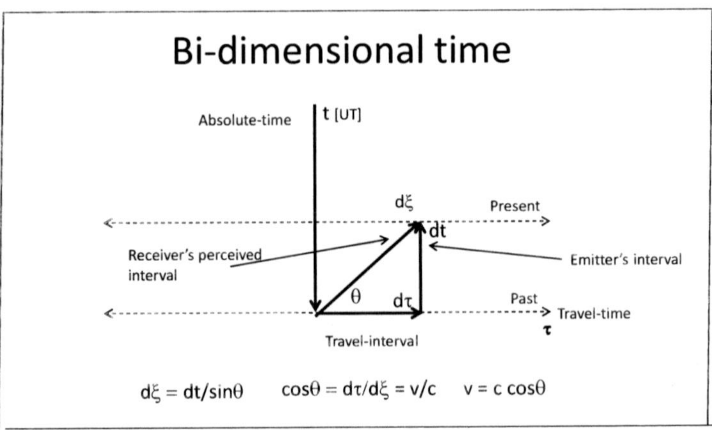



**Figure I**

If we define the unit vectors for the axis 't' and 'τ', as '$t^o$' and '$τ^o$' respectively, then:

$$\Delta\zeta = \Delta t\ t^o + \Delta\tau\ \tau^o \tag{2-2}$$

$$\cos\theta = \Delta\tau\ /\ \Delta\zeta \tag{2-3}$$

$$\Delta t = \Delta\zeta\ \sin\theta \tag{2-4}$$

$$\Delta\zeta = \Delta t\ /\ (1 - \cos^2\theta)^{1/2} \tag{2-5}$$

When interpreting a description of a phenomenon on a time coordinate system, the analyst should always account for the permanent vertical shift of the horizontal reference axis, even when $d\tau = 0$.

## Light

Currently, light is described as an electromagnetic wave travelling in space representing an oscillating energy "field" which breaks up in discrete particles when interchanging energy. As previously explained, for the TH light is also an abstract entity describing a group of individual particles [13] known as photons. Consequently, the term "LIGHT" does not describe a primary physical entity, is the assigned denomination to a mental abstraction of the brightness acquired by illuminated objects; an entelechy. Of course we could use mathematical techniques such as Maxwell's equations describing luminous phenomena when describing the effect of millions of particles structurally generated and orderly traveling together, so we can describe it as a "wave" propagating through space; but this mental abstraction albeit correctly described by mathematics, does not exist as a substantial natural entity, it simply represents the behavior of an organized group of individual photons. Of course, objectively this entelechy is helpful when investigating macroscopically natural phenomena.

A mathematical analysis of a group is always statistic, and deals with average values; but even precise statistics does not provide information about individual interaction; for instance, statistics could be very precise about predicting how many girls will be born in a city, but cannot predict the sex of a child from a particular couple, neither could be useful for analyzing the biology of fertilization.

## Speed of light

One of the most fallacious concepts in physics is the "speed of light" 'c'. The first misleading assertion is apply the concept of velocity to an

---

[13] The Theory of Hylomorphism applies the term "particle" indistinctively to matter and photons, because both are individual Hyle entities with only formal differences.

abstract entity, "light"; apart from using the term "speed", that in English language disregards the directional quality of traveled "distance" from source to destination; this vague and imprecise description of the phenomenon of natural communication considers the object "light" travelling through space, so applying to it the classic definition of speed as the "distance traveled per unit time." It is improper apply this definition to the same object that nature uses for establishing "remoteness" in a spatial environment, because it is obviously incorrect define a term by using the same subject to establish a definition; this error vanishes in the TH proposal, because it considers 'c' as a transformation factor.

Being "light" an entelechy describing a large group of travelling photons, it is reasonable to assume that all photons increase the distance to a reference point at the same rate [14].

Let's suppose we try to find experimentally the value of 'c'; it is relatively simple to measure signal travel time, but trying to experimentally measure a "distance" from an inaccessible location, without using geometric tools implies an experimental paradox; that is, in order to calculate the distance from a remote location, the signal speed must be previously known. When we want to find the distance of an inaccessible object by using only natural signals, we have to multiply 'c' by the measured travel time that took a signal in going from the emitter to the receiver. Moreover; metric "distance" is not a natural parameter, the metric unit is conventionally adopted and arbitrarily established, either by the "Meter in Paris" or the "Imperial Yard in London", whose relationship with the signal travel-time can be

---

[14] Whenever this study is forced by language semantics to refer sometimes to signal "speed", this "speed" will always be referred to the position of the emitter at the time of emission. This concept evidences that does not matter where or how it is measured, the "speed of light" will be the same since the experimenter is always measuring a universal conversion factor that translates the travel-time of a signal to distance in length units.

found by performing a controlled experiment registering the time it takes light travel a preset distance. In conclusion, what we know as the signal speed is just a conversion factor that changes travel-time to "length", mainly to fulfill language semantics.

In accordance to the explained facts, the Theory of Hylomorphism considers 'c' as a constant "transformation factor" used for transforming delay times to metric units; this transformation to distances, is justified explaining physical phenomena in an understandable manner to human perception/language and make it compatible with the concepts of classical physics [15].

Considering that delays are equivalent to the photon elapsed travel-time from the emitter to the receiver, and that this travel-time is independent of the position, the transformation factor 'c' must be a universal constant, independent of emitter properties, position, or direction; setting a "space" that is "homogeneous, continuous and isotropic", derived from the same properties from the parameter originating the concept of "space" as the entity generated by three dimensions of travel-time.

**<u>The frame of reference</u>**

The receiving particle is always informed and influenced by the characteristics of the absorbed signal; this information is defined by the emitter, which may be changing its properties like charge, mass or relative position; so, the gravitational forces and "metrix" at the receiver's location may change. Consequently, fixing a reference system at some point in space other than the receiver's location would

---

[15] Classical optics studies in physics, consider that 'c' changes within some transparent media. the Theory of Hylomorphism explains that it is due to the subjective perception of the phenomenon, derived from observer's remote position in an almost flat space, seeing a signal travelling through a denser "metrix" within the material.

mean assuming future properties and positions from all objects around it; this is why an intermediate point between emitters and receivers does not have any meaning unless the whole universe remains stable. Not even the emitter knows if the signal it sent arrived or will ever arrive to a certain particle, so its position can neither be used as a reference.

Thus, the only possible reference is consider an always stationary receiver (the observer) while the whole universe moves and changes. Moreover, it is impossible for any particle mark its previous position in an external system of in a "flat space" when "the rest of the universe is permanently moving around it." Curiously, this assertion returns physics fundamentals to pre-Galilean concepts. Large masses emitting signals continuously set properties of their surrounding environment, as a resemblance of stable "space positions", but they will always depend on the unstable influence from neighboring objects. Therefore, it is not precise studying physical phenomena by using various "frames of reference in a flat space"; when doing so, one is intrinsically assuming the stability of the surrounding environment; although, we should concede that most gravitational environments are stable enough for applying spatial concepts as conceived in classical physics.

Considering that space is a contingent entity, we can compare it with the conventional concept of "field" whose physical properties are established by the characteristics of the information carried by signals crossing that particular point, which in turn depend on the properties of the emitters in the vicinity. Under this perspective the Theory of Hylomorphism defines "sub-spaces" as entities with specific properties defined by the emitter; like the gravitational sub-space G, or the electric sub-space E [16] ; as it will be shown, both always overlap.

---

[16] The gravitational sub-space G is the one that defines distances while the electric sub-space E applies a force on a charge occupying that particular position.

## Coordinate system

The following analysis adopts the conventional mathematics representation used by theoretical physics describing "space" geometry. Since the T.H. proposal considers space as the contingent blend of velocity fields, its description is a simplification applicable to stable environments only.

**Newtonian classical mechanics** locates an object with a spatial position vector "r" whose magnitude and direction are functions of time. This vector has the three Cartesian dimensions of space whose unit coordinate vectors are ($x^o$; $y^o$; $z^o$). The instantaneous velocity [v = dr / dt] is found as the time derivative of the position.

$$r(t) = x(t)\ x^o + y(t)\ y^o + z(t)\ z^o \qquad (2\text{-}6)$$

$$v_x = dx(t) / dt \qquad (2\text{-}7)$$

Suppose that an object moves at a speed 'w', relative to a coordinate system that is moving at speed 'v' in relation to the observer, then the object speed 'u' relative to the observer is (u = v + w) known as Galilean transformation.

**The Theory of Relativity** (TR) is born when the Newtonian postulate of absolute space and time collapsed, after it was applied to light displacement; Michelson-Morley, and other experiments confirmed the constancy of the speed of light 'c', independent from the position, source velocity and observer's motion. In response to such situation it was acceptable that any change in spatial position should imply resizing the unitary time interval and the spatial distance unit (what the TH denominates "metrix" [17]); therefore demanding find invariant

---

[17] The concept of "metrix" in this book refers to the ratio between: the distance actually travelled by the signal per unit of time and the distance travelled in an imaginary flat space. In Theory of Hylomorphism terms, it corresponds to the increase of travel-time per unit of time, which is established by the signals crossing that point.

relations between detached Euclidean rigid coordinate systems in order to maintain the universality of the laws of physics. This need generated that physicists investigating the TR got accustomed to a profuse use of mathematical tools for investigating physic phenomena; a methodology that became currently essential in today's "scientific culture."

The Theory of Relativity solved the problem by establishing a space-time continuum with a coordinate system centered on the observer (x, y, z, t), satisfying the supreme condition: to keep 'c' constant. The same restrictions should apply to other inertial system (x *, y *, z *, t *) traveling away from the observer at a constant speed 'v':

$$x^2 + y^2 + z^2 = c^2\, t^2 \qquad (2\text{-}8)$$

$$x^{*2} + y^{*2} + z^{*2} = c^2\, t^{*2} \qquad (2\text{-}9)$$

Lorentz developed a transformation algorithm relating both systems as shown in the following equations:

$$x^* = \beta\,(x - v\, t) \qquad (2\text{-}10)$$

$$t^* = \beta\,(t - v\, x/c^2 \qquad (2\text{-}11)$$

$$\beta = 1/\,(1 - v^2/c^2)^{1/2} \qquad (2\text{-}12)$$

Consequently, the observed space-time intervals ($\Delta x$; $\Delta t$), from two events on the moving system ($\Delta x^*$; $\Delta t^*$) will be given by:

$$\Delta x = \beta\, \Delta x^* \qquad (2\text{-}13)$$

$$\Delta t = \beta\, \Delta t^* \qquad (2\text{-}14)$$

Under these conditions, Minkowsky proposed a four dimensions Euclidean coordinate system which kept space-time invariance by introducing time as an imaginary component of distance; this approach restricts observed phenomena remaining within a space-time observation cone.  Under this approach, the position vector 'r' for an event is:

$$r(x,y,z,t) = x(t)\ \mathbf{x}^{o} + y(t)\ \mathbf{y}^{o} + z(t)\ \mathbf{z}^{o} + i\,c\,t\,\mathbf{t}^{o} \qquad (2\text{-}15)$$

The magnitude of the position tensor will then be:

$$r^2 = x^2 + y^2 + z^2 - c^2\,t^2$$

Or: $$\qquad\qquad r^2 = x^2 + y^2 + z^2 + (i\,c\,t)^2 \qquad\qquad (2\text{-}16)$$

These equations place the origin of the coordinate system at observer's location, ignoring methodically that relative "distance" to the observed object is a secondary parameter in physics.  There is another problem generated by squaring the terms when applying the Lorentz transform; the space and time intervals give similar results for positive or negative relative velocities.

A possible solution avoid using "space" as the reference frame variable; considering that we assume be observing remote events, we could use 'c' as a constant scalar magnitude; then we can specify the "remoteness" of objects by the time 'τ' that takes a signal arrive from the emitter; these times are found by dividing 'r' by 'c' [18]; thus for each coordinate we obtain:

$$\tau_x = x(t)\ /\ c \qquad\qquad (2\text{-}17)$$

The time coordinate 't', replaces the term (i t) in **equation (2-15)**; then all 'τ' <u>are independent from universal time 't'</u>, emitter velocity 'v', or any other variable; 'τ' just depends on the "remoteness" of the emitter to the receiver; then 'r' becomes a position vector in an only

---

[18] This operation is strictly valid when the travel time magnitude of a signal travelling on the same dimension of the distance traveled at constant "speed" 'c'; astronomers use this scheme to locate stellar objects.

time environment, described by a "time coordinate system" freed from any restriction set by 'c'.

$$r(\tau_x, \tau_y, \tau_z, t) = \tau_x \, \mathbf{\tau_x}^o + \tau_y \, \mathbf{\tau_y}^o + \tau_z \, \mathbf{\tau_z}^o + t \, \mathbf{t}^o \qquad (2\text{-}18)$$

The only difference between equation (2-18) and equation (2-15) is the imaginary term, which was artificially introduced to mathematically describe the vector quality of equation (2-17).

**The Theory of Hylomorphism** (T.H.); based on the individuality of photons, places natural phenomena within a Cartesian coordinate system of four time dimensions. It can be written in tensor notation, but considering that only the absolute-time dimension ($t = \tau_4$) is irreversible, it can be written as:

$$r(\tau_i, t) = \tau_i \qquad (i = 1, 2, 3, 4) \qquad (2\text{-}19)$$

The absolute-time 't' elapses autonomously of any event; thereby, its origin will always be in the present; any previous event is past, and any future event is contingent. As for the "Travel-Time" coordinates '$\tau$', their increments could be positive or negative. The scale of the travel-time coordinate can be non linear in relation to absolute-time scale, without affecting its independence. Considering that '$\tau$' and 't' are independent variables; perceived time intervals are always the vector addition between perpendicular time magnitudes; this property demands keeping Euclidean transformation algorithms from TR, but always observing the absolute-time independence from each travel-time dimension.

The proposed conversion from the "spatial distances" to the "travel time" from the signals, is not a trivial matter because frees physics from the subjective human need referring nature to a space environment. Additionally it clearly establishes the independence between travel times, which are function of "remoteness" of the observed object on a certain dimension, and re establishes the sovereignty of "Absolute-Time" (AT) that governs the universe. By establishing solely time-coordinates, we do not need referring to "distances" to locate an event, neither even include the speed of light in the equations; the only consequence is that the conventional concepts of distance, speed, acceleration, and others should be adjusted in order to fit this natural reality.

Does T.H. perspective reflects what nature is? YES of course; it is experimentally verified that the communication is not instantaneous and that those responsible for carrying signals are photons that always spend some time before reaching the receiver; and that travel-time is proportional to the "remoteness" between transmitter and receiver. Our natural reaction to this proposal could be one of distrust, because we are talking about one of the most important parameters of physics, "the distance", being the landmark for the size of things, the height, the depth, etceteras. The concepts of distance and spatial location are imbued deeply within our human minds since they reflect our everyday experience, consequently they institute the semantic of our language; it is very difficult speaking without including the word "distance" in the phrase by using the various names that denote position and size. Considering the shocking effect on our human perception, the THEORY OF HYLOMORPHISM proposal may raise similar reactions to Galilean proposal, because in both cases somebody may say: *even a farmer knows that space exists and the sun rotates around the earth.*

It would sound rather artificial saying that you are reading a text that is 30 nanoseconds from your eyes; for the phrase being more acceptable we can add the term "light"; but even so, saying that our that neighbor lives at 9 microseconds-light sounds very strange; though it is more precise and appropriate from a scientific perspective. When it comes to long distances this reference is no longer so shocking; so that, astronomers measure remoteness of celestial objects by the travel time of their signal; this is not new, since our ancestors spoke in the same terms by saying that the nearby village was two days away, by also assuming the constancy in the speed of walkers.

We must remember that the properties of time coordinates are:

- **"Absolute-Time"** 't'. It is the universal absolute reference elapsing synchronously in all particles of the universe. Absolute time is continuous and irreversible, with an ever growing past and a contingent future.
- **"Travel-Time"** 'τ' is the time it takes the signal travel from the emitter to the receiver. It is three-dimensional, isotropic, homogeneous, continuous and <u>reversible</u>. Travel-Time interval is positive when receding or negative when approaching. As mentioned before, the "travel-Time" 3D space varies its metrix as it gets compressed by the presence of matter.

It is important to mention that the TH proposal has deep conceptual differences with the "emission theory" proposed one century ago.


## Spatial reference and straight line

The signal permanent increase in travel-time is always referred to the position of emission, effectively **<u>setting a reference point in space</u>**, from which <u>it creates a potential minimum distance geometric path as its travel-time increases</u>. As an independent entity, a photon adds travel time with respect to its starting "location", dismissing the emitter as reference after that; for instance, in conventional language semantics we could say: "the total travel time of emitted photons already traveling away, is independent from any change of location of the emitter, or even of its existence"; the emitter could be blown away by a burst of gamma rays but the photons previously emitted keeps adding travel-time to the position where it was emitted.

The geometry of the path that follows the signal also **<u>defines the geometry of the "straight line" in its own space</u>**. In this way, nature may institute a "straight" space or a "curved" space; as shown later, matter particles build their own independent circular space.

## Velocity

The classical concept of velocity, possibly born from measuring the time spend by a man walking a given distance, defined as the "speed", always relative to the rigid ground on which he walks. This definition, when generalized, contains a serious epistemological defect, caused by our human perspective about nature, since it implies that the observer must know beforehand the traveled "distance" of a remote moving object, not considering that this spatial distance is a secondary parameter impossible to be known remotely, and that it depends on the travel-time and angle of arrival of the signals that were sent at the initial and at final position of the mobile. The only place where the measures of time and space are coincident is when the mobile and the observer are together, so sharing the origin of a coordinate system; otherwise, any remote phenomenon happened in the past and its observation is always influenced by the travel-time of the signal.

In order not to unnecessarily complicate this basic analysis, we will assume that the distance traveled by a moving object during the observation interval is on a particular travel-time dimension, and that the emitter does not communicate any gravitational effect.

Under these conditions, the concept of speed ($v = ds / dt$) should reflect classical mechanics definition; i.e. the traveled distance 's' on the travel-time axis per unit of absolute-time 't'; consequently, under the perspective of Theory of Hylomorphism, the classical definition could be declared as: "***The increase/decrease of the travel time of the signals emitted by the observed object at the beginning and at the end of its trip, per unit of absolute time.***"

This definition differs substantially from the Newtonian and relativity concept of velocity, as they are geometrical (mathematical) definitions derived under the assumption that space is a substantial natural parameter; both theories establish their own geometry [19] in order to calculate distances and relate them with their respective vision of time intervals. But nature does not work like that; the relative speed between emitter and receiver is a parameter communicated by the received signal; it only acquires the quality of "velocity vector" when framed in a geometric stable space generated by a stable matter distribution within it.   The concept of velocity in Theory of Hylomorphism is explained in detail in next chapters[20].

The next paragraph analyzes the meaning of "velocity" by using the fundamental assumptions of the Theory of Relativity.

### Lorentz transformation

Let us analyze which is the time reference used by the Theory of Relativity defining velocity 'v'. Lorentz transformation establishes a coefficient 'β' that converts space-time intervals between two reference systems, one at the observer position and the other on the observed object moving at a constant velocity 'v'.

---

[19] Geometry is a branch of mathematics studying the spatial relations between distances on a three dimensional environment; but this approach does not always is applicable to natural phenomena. Mathematics is a science that applies logic rules to algebraic description of abstract entities which not always describe natural reality.

[20] The TH describes the phenomenon as; photon informs '$\Delta\tau$' to the receiver locally where ($\Delta\xi = \Delta t$), this definition should accommodate the real world while observing Theory of Relativity notation. Then the conventional velocity definition ($v = c\, \Delta\tau / \Delta\xi$), is compatible with the local measurement perceived by the receiver as the asymmetry from the absorbed photon calculated by the Doppler Effect equation, i.e. [$v = c\, (T - T\,^*) / T$]; where 'T *' is the period and 'T' the travel-time length of the photon.

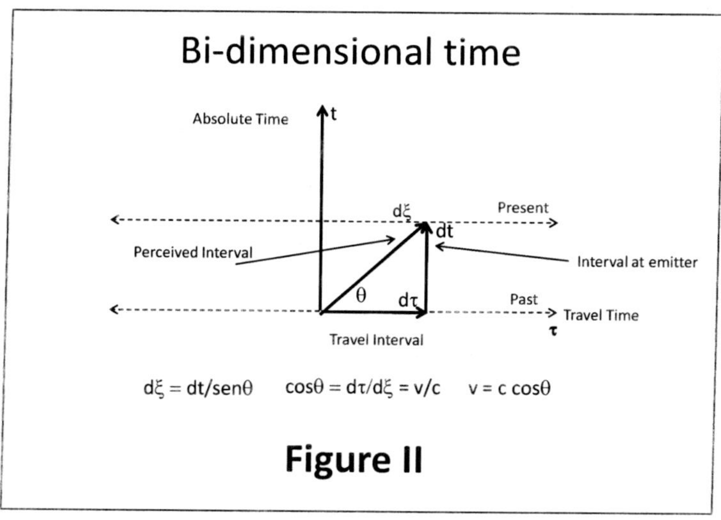



**Figure II**

The classical (Newtonian) definition of velocity is "the traveled "distance" 'ds' per unit of absolute time; where 'ds' is found by transforming the travel time units 'dτ' in spatial units (ds = c dτ), so the classical velocity definition in terms of the Theory of Hylomorphism is:

$$v = c \, d\tau \, / \, dt \qquad (2\text{-}20)$$

In accordance to figure I:  $\cos\theta = d\tau / d\varsigma$  (2–21)

And, in accordance to Lorentz equation:

$$\cos\theta = v \, / \, c \qquad (2\text{-}22)$$

Therefore, Lorentz equations implicitly define distance as:

$$ds = c \, d\tau = v \, d\varsigma \qquad (2\text{-}23)$$

This equation no longer reflects the classical Newtonian definition of velocity.

Consequently, the effective Theory of Relativity velocity definition is "traveled distance per unit of perceived time"

$$v = c \, d\tau \, / \, d\zeta \qquad\qquad (2\text{-}24)$$

By applying this Theory of Relativity definition to the figure above we easily find the Lorentz transformation equations as follows.

If: $\qquad\qquad\qquad\qquad \cos \theta = v \, / \, c$

Then; $\qquad\qquad\qquad\quad v = c \, d\tau \, / \, d\zeta \qquad\qquad (2\text{-}25)$

$$\sin \theta = dt \, / \, d\zeta \qquad\qquad (2\text{-}26)$$

By defining $\beta$ as:

$$\beta = 1 \, / \sin \theta = 1 \, / \, [1 - (\cos \theta)^2]^{1/2} \qquad (2\text{-}27)$$

Then: $\qquad\qquad\qquad \beta = 1 \, / \, [1 - (v \, / \, c)^2]^{1/2} \qquad (2\text{-}28)$

Therefore, time transformation is given by:

$$D\zeta = \beta \, dt \qquad\qquad (2\text{-}29)$$

This result confirms that the **Theory of Relativity implicitly considers 'd$\zeta$' as the time interval defining velocity.**

Under this definition the value of observed velocity from a distant object travelling at constant velocity on a linear trajectory on two travel-time dimensions will depend on the angle '$\alpha$' of its displacement with the radial receiver-emitter line; the radial component of displacement is ($d\tau_r = ds \cos \alpha \, / \, c$) defines the value of 'd$\zeta$' and therefore ($\cos \theta = v_r \, / \, c$), while the lateral (tangential) displacement ($d\tau_t$ = ds sin $\alpha$ / c) will not be influenced by the signal delay[21].

Calculating acceleration under this basis may lead to further subjectivity applied to the observations of natural phenomena.

Lorentz spatial transformation (space expansion) is meaningless for the Theory of Hylomorphism, but it is interesting exploring what is the concept of distance in accordance to Lorentz spatial transformation; this may be found by redefining distance as (ds' = v dt = c d$\zeta$) or (v = c d$\zeta$ / dt); then:

$$ds = c\, d\zeta = \beta\, c\, dt = \beta\, ds' \qquad (2\text{-}30)$$

The previous analysis shows that the Theory of Relativity effectively define a concept of time by blending time coordinates as the result of considering "distance" an independent variable and "space" a substantial entity; moreover, by operating on time as being a space-like coordinate it subjugates time to the same category of "space", disregarding its substantial condition.

It is important realize that alternative velocity definitions as (v = c d$\tau$ / d$\zeta$) or (v = c d$\zeta$ / dt) may generate additional conceptual imprecision; when calculating acceleration we could apply one of many definitions as (¿dv/d$\zeta$?; ¿dv/dt?; ¿dv/d$\tau$?); ambiguity gets worse in higher order derivatives. Considering above arguments, it is not surprising that time related results under Theory of Relativity perspective may wind up in paradoxes.

---

[21] The change in radial position is observed as an increase/decrease of signal travel time '$\Delta\tau$' and the change in angular position '$\Delta\phi$' represents a "flat space distance" ($\Delta\phi\,\tau$), so that the total distance traveled by the observed object is the vector addition of the two orthogonal components $\{[(\Delta\tau)^2 + (\tau\,\Delta\phi)^2]^{1/2}]\}$.

The "subjective" definition of speed effectively adopted by the Theory of Relativity leads observer erroneously concluding that "time does not pass for the emitter" ($\Delta t = \infty$) when it changes its position at such a rate that signal delay time changes as fast as absolute time ($\Delta \tau = \Delta \zeta$), (conventionally stated as $v = c$; $\cos \theta = 1$ and $\theta = 0°$).

## Travel-time dilatation analysis

To clarify the differences between the Theory of Hylomorphism and the TR, herewith follows a brief analysis on time/space dilatation as seen under the postulates of TH.

By using geometrical methods it is easily perceived that the hypotenuse is larger than the sides of the triangle by simply observing it; on the temporal coordinate system plane, emitter, receiver, and signal travel-time are measured by using the same universal time unit and their events occurring at a time given by the "universal hour" elapsing simultaneously for all of them. So their locations in a time coordinate system are not intuitive to our "spatially oriented mind." The vertical axis 't' shows time given by the emitter's clock; horizontal axis '$\tau$' is the travel-time between the emitter and the receiver, and the path length is the elapsed time as given by the receiver's clock.

The temporal plane we refer in **Figure III** consists of two orthogonal axes corresponding to the two independent variables whose unit vectors are '$\tau^o$' and '$t^o$'; the first represents the travel-time unit vector and the second the emitter's clock absolute-time unit vector. Considering all clocks in the universe show the same time and advance at the same rate, wherever they are, the elapsed magnitude of both times is the same.

The receiver keeps track of the historical sequence of signals sent from the emitter at intervals 'dt', perceiving them at time intervals 'd$\zeta$' as it is an influence induced by the change of position 'd$\tau$'. Total time '$\zeta$' is the addition of 'd$\zeta$' resulting in a "time path length." Vertical time axis 'ζ' is drawn in order to compare the "length" of the path sharing scale with axis 't'.

In general, any trajectory is longer than the emitter's elapsed absolute-time, whether it is approaching or going away; i.e., the travel-time interval always adds delay to absolute-time. The emitter can move away or come closer, but it will always be on the line of "present" which rises permanently. The angle between the tangent and the horizontal is 'θ'.

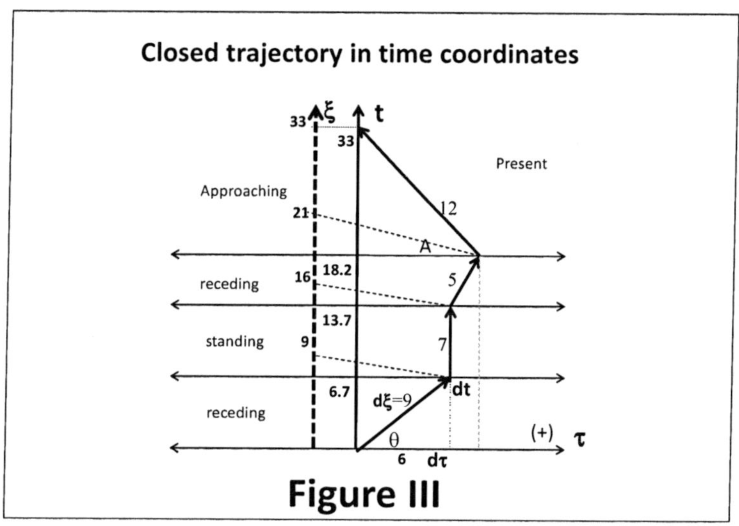



**Figure III**

**Figure III** represents the sequence of events of an emitter as it moves away and returns in a "delay-time closed path": the vector 'dζ' represents the receiver's <u>perceived</u> time interval for every new event with respect to the previous one; consequently, its addition represents the total time spent travelling along the trajectory. It is impossible to have a completely "closed space trajectory" as simplified by classical theories; absolute-time never stops, one particle may travel away and come back to the same "place", as shown in **Figure III**, but it cannot return to the past. At this point it is clear for the reader how Theory of Relativity confuses both time dimensions, attributing to absolute-time qualities of travel-time.

The accumulated travel-time in a closed path converts into an unrecoverable past; longer trips produce larger elapsed time. The Lorentz transformation states that it will always be a perceived time dilatation due to the squared term, no matter whether objects are coming closer or going farther away, but it is very easy recognize that the perceived absolute-time difference between the clocks of the emitter and the receiver, is reversible and vanishes for both clocks at the end of the lap.

It does not matter, whether the emitter and receiver are together or far away from each other, their clocks will mark the same time ($|\zeta_1| = |t_1|$); but as they keep separating, both read that the other's clock is running slower, thus increasing the perceived time difference; when they approach each other each perceives that the other's clock is advancing faster as the universal time difference decreases. Nature behaves as described; physicists use it when observing the frequency shift of spectral signatures for objects approaching (red shift) and moving away (blue shift).

In order to find the relation between perceived times and the velocity at which the emitter approaches or moves apart, we must find the perceived universal-hour difference '$\Delta\zeta$' from the beginning to the end of the motion; we will suppose that both are separated by an initial delay '$\tau_o$'. The emitter sends a signal communicating the hour '$t_o$' [UT] as given by its clock; the receiver reads the information at '$\zeta_1$' [UT].

$$D\zeta = \zeta_2 - \zeta_1 \tag{2-31}$$

Where:
$$\zeta_1 = |t_o|\, t^o + |\tau_o|\, \tau^o \qquad (2\text{-}32)$$

If we consider the starting hour in universal-hour is zero ($t_o = 0$)

$$\zeta_1 = |\tau_o|\, \tau^o \qquad (2\text{-}33)$$

Note that the direction of '$\zeta_1$' is '+ $\tau_o$'; for the emitter it would be '- $\tau_o$'.

When the emitter moves, the receiver's time advances as the signal's travel-time "$d\tau$" increases; meanwhile, the emitter's clock advances 'dt', resulting in a perceived time equal to the vector addition of two independent times. The direction of this vector points to the right or to the left (+ $d\tau$ or -$d\tau$) as the emitter approaches or goes farther away respectively.

$$\zeta_2 = dt\, t^o + (\tau_o \pm d\tau)\, \tau^o \qquad (2\text{-}34)$$

$$D\zeta = dt\, t^o + (\tau_o \pm d\tau)\, \tau^o - |\tau_o|\, \tau^o \qquad (2\text{-}35)$$

$$D\zeta = dt\, t^o \pm d\tau\, \tau^o \qquad (2\text{-}36)$$

After several stages; $\qquad \Delta\zeta = \Sigma dt\, t^o \pm \Sigma d\tau\, \tau^o \qquad (2\text{-}37)$

$$\Delta\tau = \Delta t / \tan\theta \qquad (2\text{-}38)$$

$$|\Delta\zeta| = [(\Delta t^2 + (\Delta t/\tan\theta)^2]^{1/2} \qquad (2\text{-}39)$$

$$|\Delta\zeta| = \Delta t / \sin\theta \qquad (2\text{-}40)$$

This is the Lorentz Transformation formula; it simply confirms that the hypotenuse '$\Delta\zeta$' is always positive because the receiver clock is always ahead, and larger than '$\Delta t$' or '$\Delta\tau$'.

After returning to the starting position, the total elapsed time '$\zeta_t$' perceived by the receiver has two components; the first on the vertical axis ($t_t = \Sigma\, \Delta t$) that represents the total absolute-time, which is irreversible and second, the horizontal component ($\tau_t = \Sigma \, \Delta\tau$) that is the accumulative travel-time at the end of the trip; it can be positive, negative, or null as being the total relative delay between emitter and receiver clocks.

$$\zeta_t = \Sigma\Delta t \; t^\circ + \Sigma\Delta\tau \; \tau^\circ \qquad (3\text{-}41)$$

$$\cos\Theta = \tau_t \, / \, \zeta_t = v_m / \, c \qquad (3\text{-}42)$$

Where '$v_m$' is the average speed observed if the emitter moves directly to the final destination with an angle '$\Theta$' with '$\tau$'. A closed path means that ($\Sigma \, \Delta\tau = 0$); consequently both clocks when together will show the same time, as they always do; this fact can be confirmed by an equidistant third observer perceiving that both clocks were always marking the same time, although delayed as compared with the time given by observer's clock and vice versa.

This clarifies the conceptual differences about time between the TH and the TR. The Theory of Relativity supposes a single time dimension related to space by the invariance of 'c', while Theory of Hylomorphism considers 'c' as a transformation coefficient "_applicable to the travel-time coordinates_" when converting to metric units.

For TH, all observed changes in length, mass, and time are just subjective perceptions from the receiver's perspective; for example, the forces applied remotely in order to accelerate a particle undergo a huge increase when the particles escaping from the source reach speeds equivalent to the signals that "carry the force" pushing them; additionally, the increase in mass is perceived because a smaller quantity of "force carrying" particles will arrive on time at the pushed particle as its speed increases; all this perceptions are absolutely subjective perception generated by the finite signal's speed.

## Chapter III

## Time symmetry: the Fundamental Natural Law

The two ingredients building the whole universe and its content are: the time, and the signals communicating information among particles. Time elapses independently from any other natural parameter; and the signals are time structures communicating objects residing in distant locations.

The concept of "present" is individual and implies simultaneity, so it can be applied only to entities being at the same location. The present time, as a universal concept, is an entelechy which assumes that the observer has a "divine" capability: to instantly perceive everything that happens everywhere. The present is always elapsing at a given pace, permanently marking the line separating local unchangeable perception of past events since any distant phenomena happening "now", anywhere, may only influence the receiver in the future.

The TH represents our intuitive concept of time as the variable representing the absolute–time (AT) dimension. It is the only natural factor elapsing independently from any other natural parameter; its only self-reference is the moment that started everything. We normally use "the present" as reference, although sometimes it is useful use some reference in the past; or chose marking the start when a future time becomes being "the present."

AT is a one-dimensional parameter for our particular universe. Multiple independent AT references would mean that natural phenomena may not have a unique reference, causing that universal principle of causality, as observed in nature, will not be possible. Under this scenery, emitters will have diverse time unit reference, so receivers will be unable in decoding information; additionally the signal "speed" would be chaotic.

The TT magnitude is independent from AT, but it is measured using the same universal time unit; the second. The TT always starts at zero at the emission point and has <u>the potential defining a distance</u>. This parameter is also contingent; it becomes real only after the emitted signal arrives to a receiver. So the "distance" separating two points in "space" is a conditional entity, and valid only for every specific time that a signal is received; reception is the only way establishing "remoteness"; furthermore, the straight-line geometry is always defined by the signal's 3D trajectory.

The TT of any signal is always independent from AT; because its origin is at the emission point. The virtual straight-line distance in a three dimensional preset environment is defined only when there are millions of signals following similar paths; remember that even the concept of path also requires the permanent interaction of signal along it.

Consequently, time is bi-dimensional by nature having the AT dimension and the signal generated TT dimension. The 3D subset of TT is a secondary feature set by nature accommodating complex structures containing large conglomerates of emitters sending signal to all possible directions; and so, produce a tridimensional "space." Unless these premises become certain, it is not absolutely proper stating that a signal from a distant source "moves through space" since the past most recent interaction defined the relative speed from one emitter and the next signal from a different emitter will give another value, so the statistic blend of the received data from all sort of signals will statistically define the geometry of surrounding environment as perceived by a particular receiver; with no interaction, there is not possible way defining the relative speed or the "remoteness."

The TT of a signal sets the fastest rate at which distance from a given emitter can be <u>generated</u>. Although the TT scales may be not linear.

## The space-time symmetry

In the following analysis it will be shown that the cornerstone, on which the universe is erected, is the fundamental law establishing that time intervals elapse at the same pace in every Hyle structure in all time dimensions, at any point in the universe. This rule effectively sets the absolute time pace as the sole universal time unit for all entities residing in the material universe.

This seemingly trivial premise should be appraised in its whole profoundness; it not only sets a universal time unit, but also founds the base for raising "the spatial metrix." [22]

The AT intervals run independently of any other natural parameter; on the other hand, the TT intervals in "space" mean "the rate at which TT increases", as set by a particular photon. The "remoteness" is not a pre-established parameter; it is the result of the total time travelled by a particular signal, the TT smaller discrete interval that a signal increases is one "length" 'T' per period 'T*' (in AT). A distant observer perceives this ratio as the apparent signal speed in that particular sub-space, conventionally known as "the speed of light."

The natural tendency to share the same time interval in the particle time dimensions denotes that symmetry is the final objective for a Hyle structure; this tendency may be defined by saying:

---

[22] The TH defines "metrix" as the ratio between travel-time intervals and absolute-time intervals within particle structures. The same concept is applied to abstract entities as "space" and "fields."

## _Every segment of a Hyle structure tends to increase travel-time at the same pace than absolute-time._

Acknowledging the substantial nature of Hyle structures, there must be the tendency of every segment of Hyle structures <u>reaching internal symmetry</u>; when the structure is compressed or expanded[23], it will recover equilibrium, like a common spring.  This tendency could be stated as:

$$\delta\tau \rightarrow \delta t \qquad\qquad (3\text{–}1)$$

Or it can be written as a ratio:

$$\xi = \delta\tau / \delta t \rightarrow 1 \qquad\qquad (3\text{–}2)$$

This ratio establishes the final status of any particle; The "LAW OF TIME SYMMETRY", or simply the "law of symmetry", imposes the **"<u>universal tendency of Hyle structures having the same travel-time interval in all dimensions</u>"**; when this law is applied to photons shows the trend of photons for flattening the space they define as they travel. It also establishes the elasticity of any Hyle structure, allowing Hyle build matter particles as independent <u>elastic structures</u>.

The ratio between travel-time dimensions and absolute-time '$\xi$', is a non-dimensional parameter which also be equal to one ($\xi = 1$) in all structures. The TH denominates this parameter as the "metrix" of an Hyle structure which also may be assigned to the metrix of a given spatial location.

---

[23] The dimension of a Hyle structure can be affected by external influences as: mechanical stress or by changing the spatial curvature.

The ratio between the travelled "distance" and time, applied to a signal is conventionally known as "light speed"; then, 'ξ' may also represent, in conventional terms, <u>the relative speed of the signal</u> perceived by a distant observer compared to the speed of a signal in a flat space.

The difference between the conventional symbol 'c' representing the "speed of light", and the conversion factor 'c' in the TH should be clarified in the following example: let's imagine that a distant observer could "see" the compressed ($\xi < 1$) photons "a" travelling perpendicularly to the line of sight, and compare them with symmetric ($\xi = 1$) photons "b" travelling parallel; the "a" photons will be perceived as travelling slower. The observer may describe the results <u>in classical semantics</u> saying that *"the photons "a" travel slower than 'c'; or he could say that the space in which are the "a" photons is curved"*; the same result can be described under the TH terms, saying that *"the photon "a" metrix is smaller than one"* ($\xi < 1$). Local experimenters close to the photons, will perceive both signals ("a" and "b") travelling at light speed ($c = 1$); and only if they specify its value in conventional speed units, they should apply the convenient conversion factor 'c' that correspond to the system of units used.

As it will be demonstrated in the following chapters, this surprisingly simple and absolutely rational law is the cornerstone which makes possible deduce the most basic law of physics; establishing gravity by "curving space", it also defines force and elasticity, and sets the rules for composing atoms with diverse particles; in short, it rules the way that the universe is built.

**<u>The Hyle: a four dimensional substance</u>**

Another fundamental premise for building a multidimensional universe is that the entities residing on it must be constituted by a substance with the potential to occupy all those dimensions; stating this concept under the TH perspective, <u>the Hyle structures must be potentially four dimensional</u>.

The insightful meaning of this assertion establishes that: Hyle, by occupying the absolute-time dimension evolves permanently with time; occupy all travel time dimension means that it is a structure containing vectors pointing to all spatial dimensions along it. Consequently it is a "three dimensional vector structure moving permanently."

The Hyle equation proposed later in this chapter complies with this basic demand.

## Hyle, the only substance

It would not be clever building a universe with many ingredients, photons, matter, force carriers, gravity, electric waves, magnetism, time, fields, space, etc.; consequently, it is reasonable assume that all this parameters represent the form taken or the effect caused by a common substance; consequently the entire physical universe must contain a single substance constituting various entities by changing its form and behavior. The combined dosage of several substances building a particle would not facilitate the formation of diversity, because this approach automatically increases the implicit restrictions that define the particular qualities of every component[24].

Accordingly, TH postulates that it takes only one substance with two basic components, time and form, to assemble a unique structure governed by some simple laws, suffices for acquiring the full range of

---

[24] Proof of this is the variety of incompatible proposals plaguing modern physics, each of which limits its scope due to the specific characteristics of each alleged substance.

shapes, properties, and structures composing everything in nature: photons, atoms, molecules, stars, and planets. Furthermore, the TH perspective is so clear that one can comprehend that the simplicity that rules nature does it so wisely that is not surprising that matter evolves to generate life forms, and creates consciousness.

The TH refers to that one universal substance as "Hyle" [25] having an essential component, time, and a specific form. This name was adopted as a deserved tribute to one of the greatest minds in human kind, Aristotle; who with an insatiable curiosity and intelligence has covered a broad spectrum of topics on philosophy, physics, politics, ethics, aesthetics and biology; the depth of his proposals was the result of the extensiveness of his interest. Profoundness is impossible to achieve when ignoring universal context.

Any proposal, satisfactory for a branch of knowledge which contradicts the principles of some other, must be false or incomplete because it implies that at least one of the two disciplines lacks of solid foundations on a particular subject. The base of all science originates in metaphysical principles that have to be respected by all branches of science, be it physics, economics, psychology, etc. In the case of physics, it is wrong validating paradoxes, or assuming undetectable materials, or to pretend the reversibility of time, or to have instant communication; I feel imperative to mention this concepts because currently, became usual propose theories which contradict basic natural laws and philosophical principles. The TH proposal is consistent with both, metaphysical and natural laws, discarding the false and the absurd as nonexistent, and therefore unreal.

---

[25] "HIΛE" in ancient Greek means raw material.

## Signals and information

Considering that signal carriers and particles are entities made with the same substance, it will be convenient to initially investigate the characteristics of photons in order to know the structure of particles.

Let's remember some basic general logic statements underlying the TH proposal:

- Nature's wisdom implies simplicity; consequently, natural basic principles must be very simple.

- It is impossible to find the initial spatial position of the universe; therefore, space is not a pre-existing natural entity.

- The observed permanent change means that time rules on nature. In every natural phenomenon, the cause always precedes to the effect, implying the irreversibility of absolute-time.

- It is reasonable to assume that for sending/receiving signals, a particle must be able to emit/store them. It could do so it by having a suitable reservoir for every type of signal, but is reasonable assume that more likely nature solved the problem without different materials or special deposits. It is more probable that nature used the same raw material as the substance forming particles, signals and containers.

- The reception of a signal reception implies its absorption; so signals, are necessarily made with the same substance as particles for producing a substantial change upon absorption, since nature demands <u>active communication</u> in order to perform as a whole. Signal absorption must produce internal structural changes in the substance of the receiver, directly related with the received information.

- A signal cannot be a continuous entity lasting indefinitely, since it would exhaust the substance of the emitter; consequently it has to be a discrete individual entity whose form defines its properties; the magnitude of its influence is achieved by the aggregate individual effect.

- To communicate complex information to distant particles, the Hyle in a signal must be an oscillating entity in perpetual motion.

- An externally induced imbalance inside a particle must cause signal emission; consequently, emission requires a previous absorption, which in turn demands a continuous communication between all objects in nature.

The TH assumes that the photon is the natural "information carrier" so its substance must be the same that the one that constitutes all particles. Let us analyze some of the particular characteristics that should have this natural information carrier.

Photons are individual travelling entities departing from the emitter so they move apart from it permanently until they find a receiver; in this way, the emitted signal created a spatial reference that is the basis to potentially define a "distance", only after it was received.

Particles always interact locally, i.e. any interaction always takes place between entities <u>being in the same place at the same time</u>; therefore, one of the interacting particles would be a photon communicating some information to the receiver about the status of a distant emitter. This information must include all the individual characteristics from the emitter as: type, size, mass, distance, charge and polarity.

The amplitude of the signal carrier will not suffice because it does not have the required capability to contain complex and diverse information, the Hyle must be a multidimensional oscillatory entity; in other words, it must contain a rotating vector. Therefore is reasonable assume that signals must use their dimensional capabilities for encoding complex information; this means that they use the three travel-time dimensions and the absolute-time dimension in order to compose oscillatory structures. Only oscillating entities are capable to transmit and communicate complex information as Hyle pass through a point in space; on the contrary, static carriers could only communicate amplitude during a given period.

All signals travelling away from their emitters carry time coded information, it is mandatory have a precise time reference to synchronize and decode it; consequently; the only solution fulfilling this requirement is that all particles may accede to a universal clock, but as all of them are independent entities, then they must have their own reference internal clock <u>elapsing exactly at the same pace</u>.

Considering that signals communicate information among distant particles, and define the travelled time; their <u>path defines the minimum distance</u> in the spatial dimension connecting the emitter with the receiver.

Is important pointing-out is that in a three dimensional environment, the probability of a single signal emerging from an individual emitter arriving to one specific receiver is very low. So, nature demands a massive emitter composed by millions of individual particles sending constantly large quantity of signals in all directions; this is only possible when the particles in the emitting body are interchanging signals permanently[26].

---

[26] This is the concept of temperature; a large quantity of signals inside an object means high temperature. On the contrary, cold objects are easily influenced by external signals (as in superconductive materials)

The information is always contained within individual photons; considering that some bodies emit large quantities of photons, it is easy miss "reality" by oversimplifying the phenomenon investigating the average behavior of the group mechanics; but averages cannot explain details. Analysis of averages is a mathematical statistical procedure used for investigating massive interaction; analyze an ultimate cause; one must dismiss the average behavior and concentrate on the individual phenomenon.

Currently, physics perspective describes natural phenomena as "fields" communicating forces acting on receivers exerted by "emanations" originated by emitters. One must keep in mind that the term "field" refers to the mathematical description of a mental image that associates a great number of signal carriers whose space-time distribution was somewhat related at the time of generation; so "field equations" quantify the subsequent effect on material objects located within reach. In general, field equations are not the right tool for investigating individual interaction between particles, which is ultimately the origin of the natural phenomenon; natural interaction always involves two individual particles; the concept of "field" provides a useful tool for finding global relations. For instance: Maxwell equations cannot find the mechanics of interaction between particles and photons; gravity is a mystery when ignoring individual particle behavior.

## The Hyle

The unifying quest in physics starts when Newton finds the direct relation between mass and force, next step was given by Einstein establishing the mass-energy equivalence. The TH, regards indistinctly: mass, energy, force, momentum, as formal particle properties given by its particular multi-dimensional time structure; size, form and stability.

The signal carrier (photon), that is the basic Hyle structure, cannot be described as a punctual event in time as a single rotating vector touching only a point of the receiver's structure; it would could not communicate complex information to it; consequently, a photon has to be a rotating structure with a spatial length, a spatial shape and an angular speed.

In line with these arguments, the TH proposes that Hyle has two components: time, the universal and unique substance, and shape. Every single particle is a "discrete time entity" residing in all four time dimensions; its magnitude in the absolute time dimension 't' instituted by a period of rotation ($T^*$) and its three travel-time dimensions '$\tau_i$' given by its particular shape and its "length" (T).

The ontological ingredient is the binary magnitude of the vector '**H**'; then Hyle in a signal can be described as a <u>vector function</u> whose binary vector '**H**' takes the value of "1" or "0" (exists or does not exist in a given 4D location). This finite vector structure permanently rotating at a constant angular velocity while travelling away from the emitter; its rotation period 'T*' represents its own absolute-time reference; and the perpetual travel-time increase (from the emitter) generates [27] the metrix in that point of the travel-time dimension of its path.

Consequently, the two components of Hyle structures are:

- **Time**; which is the essence of everything. Nothing **remains** unchanged; neither **stays** in any relative position, and does not **exist** forever.

- **Hyle vector**. It is a binary vector in perpetual motion which uses the rotation for establishing moving multi-dimensional time structures, and defines the limits (length) of any structure, the potential geometric properties of the photon, and directional scale

---

[27] Any property of a point in space is just a contingent entity; it is always subject to the existence of a receiver in that point. This is the case of defining the metrix at a spatial location.

(metrix and linearity) of the specific spatial travel-time dimension its path.

Specifically, the TH proposes that all particles have the same basic oscillatory structure travelling on one spatial dimension; this structure is geometrically described as a helical continuous vector function with a length 'T', rotating at a period 'T*', comparable to a rotating helix. It must be kept in mind that the fundamental law of nature establishes that every part of this structure must be symmetric, increasing always the travel-time interval 'd$\tau$' equivalent to the elapsed absolute-time interval 'dt'; consequently its length 'T' must always tend being equal to 'T*'.

This model explains how all particles contain their individual absolute-time reference given by its period of rotation 'T*' always running synchronously with universal time. This property is indirectly established by the well known Einstein-Planck relating the energy with the frequency of the photon.

The rotating helix travels projecting a sine wave with a given frequency on all three travel-time dimensions whose amplitude depends on the angle between the displacement direction and the plane of rotation.

This page is intentionally left in blank

## Chapter IV

## Hyle structures

## Photon Hyle equation

The equation describing the structure of a particle is a discrete vector function '**H**' of finite length[28]. The photons have to be oscillatory, finite and individual objects that can be placed on a Cartesian coordinate system representing the three independent travel-time dimensions of space ($i^o$; $j^o$; $k^o$).

Let's consider the oscillating substance "Hyle" moving in the direction '$k^o$' and the vector function '**H**' describing it as a unit vector rotating at an angular velocity ($\omega_o = 2\pi / T_o{}^*$) in a plane containing '$j^o$' so the angle '$\psi$' between '$k^o$' and '$\omega_o$' builds a continuous structure whose angle with the axis '$j^o$' is:

$$\gamma = 2\pi\,(t / T_o{}^* + \tau / T_o)$$

So its equation is:

$$\mathbf{H}(h\,;\,t;\,\tau) = h\,(\operatorname{sen}\gamma\,\cos\psi\;\mathbf{i}^o + \cos\gamma\;\mathbf{j}^o + \operatorname{sen}\gamma\,\operatorname{sen}\psi\;\mathbf{k}^o)\quad(4\text{-}1)$$

Where: $h = 1$ $(0 < \tau < T)$; and $h = 0$ for all other values of '$\tau$'

Where:  - 't' is the absolute-time, AT.

---

[28] Some other model to study a particle could be raised; for instance, considering it as an oscillating fluid in a container of a certain length, which could be very useful for analyzing absorption phenomena, but it is likely that the results would be very similar to those developed in the present study.

- '$T_o^*$' is the period in AT, at which the structure rotates one full revolution.

- '$\tau$' is the travel-time TT, at which one point of the structure moves.

- '$T_o$', is the length of the particle, in TT.

- $\omega_o$ has an angle '$\psi$' with '$\mathbf{k^o}$'; ($\omega_o = 2\,\pi\,/\,T^*$).

- 'H' is the module of the rotary unit vector '$\mathbf{H}$' of the structure.

When the rotation plane is on ($\mathbf{i^o}$; $\mathbf{j^o}$); perpendicular to ($\mathbf{k^o}$), then:

$$\mathbf{H} = \mathrm{sen}\gamma\ \mathbf{i^o} + \cos\gamma\ \mathbf{j^o} \qquad (4\text{-}2)$$

The structure described by the above equations, is a single spatial dimension entity represented by a straight line that starts when ($\tau = 0$) and ends when ($\tau = T_o$) that has a unit vector spiraling on all its length '$T_o$'; the whole structure spins a complete revolution in an absolute-time period '$T_o^*$' while travelling a travel-time '$T_o$'[29].

The Hyle structure advances while rotating, resembling a screw that is threaded into the space, thus the projection on axis "$\mathbf{i^o}$", "$\mathbf{j^o}$", or any other intermediate axis perpendicular to "$\mathbf{k^o}$" is a sinusoid of unitary amplitude and a wave length ($\lambda = c\,T_o$).

These equations describe a physical entity as a continuous vector function that changes direction forming a helix of length '$T_o$' after completing a revolution in a time period '$T_o^*$'. Nevertheless, as the angle between "$\mathbf{k^o}$" and the plane of rotation may not always be $90^o$, the shape will not always be a classical helix.

---

[29] Symmetric particles have ($T = T^*$) and are said to be in equilibrium. They may lose symmetry when subject to external influences as tension or spatial metrix change.

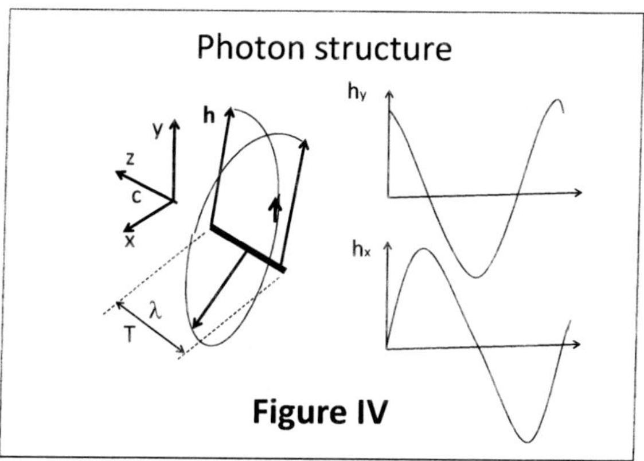



**Figure IV**

## Matter particles

The TH proposes that a material particle is described by the same equation; the only variation is that this description is adapted to the subjective perspective of an external observer perceiving it as a ring because the displacement direction "$k^o$" is always tangent to the circular path which is the straight line in the internal space of the particle. Hyle's path always defines the geometric characteristics of the space it passes through; in the case of a particle the straight line is a circle. The apparent incoherence of having a closed space, where "advancing" means always travelling on the same path, which is the key factor creating independent particles (matter) residing in their own-space, which as a circular structure can occupy a location in an external "space."

The objective of describing geometrically the shape of a material particle, or matter ring <u>as seen by an external distant observer,</u> is setting a common reference geometry that enabling the investigation of the interaction among particles as seen from the perspective of an observer residing in an ideal external flat space; natural phenomena are more easily analyzed and explained under this vision, because it is normally the perspective of a physicist observing a natural phenomena.

The general particle's equation for the variables shown in the next figure is:

$$H(t; \tau) = \sin \beta \, \mathbf{i}° + \cos \beta \cos \gamma \, \mathbf{r}° + \cos \beta \sin \gamma \, \mathbf{z}° \qquad (4\text{-}3)$$

$$\beta = 2 \pi \, \omega_o$$

$$\gamma = \arccos(\mathbf{z}° \circ \omega°); \; 0 < \gamma < \pi/2$$

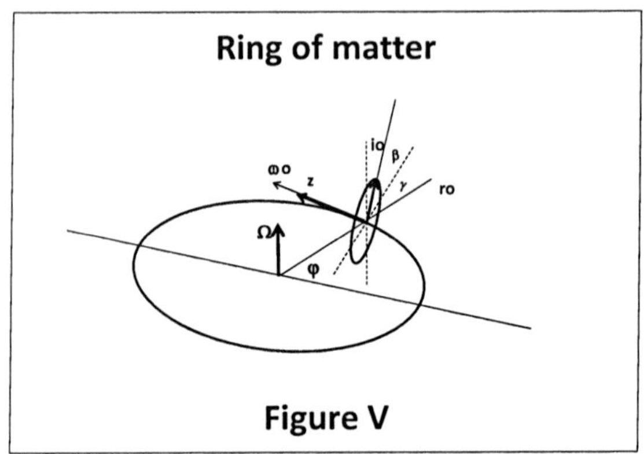


**Figure V**

'β' is the angle between 'H' and the rings plane, as it rotates with an angular frequency '$\omega_o$'; 'φ' is the angle of the plane of rotation with the ring's initial position.

The structure completes a complete rotation in a travel-time period 'T'[30], with an angular speed 'Ω' (T = 2 π / Ω)

Any stable particle has ($\omega_o = \Omega$) and ($\beta = \varphi$), so vector 'H' always points in the same relative direction in every angular position on the ring,

---

[30] In order to facilitate the interaction analysis between a photon and a particle, normally the period and length of a photon will be 'To*' and 'To' respectively; and 'T*' and 'T' for a particle.

instituting matter as a "<u>time independent structure</u>" that seems static to an external observer.

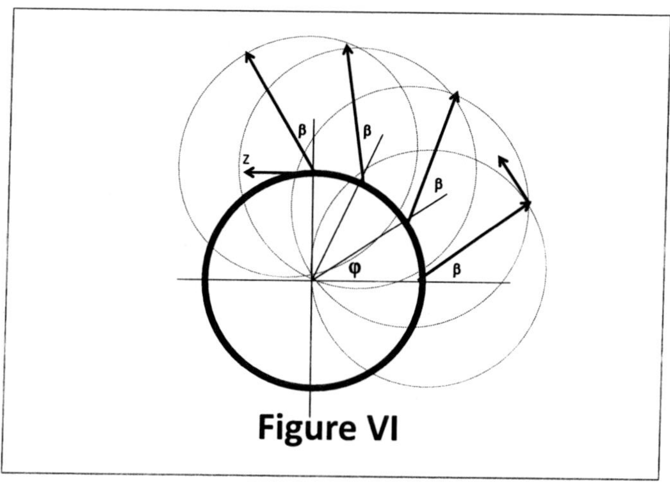

**Figure VI**

**Figure VIII** shows the equation of '**H**' when the plane of rotation of '**H**' coincides with the circular ring:

The angle '$\alpha$' between '**H**' and the radius is always ($\alpha = \beta - \varphi$):

$$H = \sin \alpha \; x^\circ + \cos \alpha \; y^\circ \qquad (4\text{-}4)$$

$$\beta = \beta_0 + \omega_0 \tau$$

$$\varphi = \Omega \tau$$

$$\alpha = \beta_0 + (\omega_0 - \Omega)\,\tau \qquad (4\text{-}5)$$

As ($\omega_0 = \Omega$); $\qquad \alpha = \beta_0 \qquad (4\text{-}6)$

When ($\omega_0 = \Omega$) $\qquad H = \sin \beta_0 \; x^\circ + \cos \beta_0 \; y^\circ \qquad (4\text{-}7)$

This equation describes a static structure having a constant unit magnitude pointing in the direction set by the phase angle '$\gamma$'. The vector function '**H**' can be described in polar coordinates as:

$$\mathbf{H} = \sin \beta_o \, \mathbf{z}^o + \cos \beta_o \, \mathbf{r}^o \qquad (4\text{-}8)$$

The radio 'R' of its circular path in metric units is: ($R = c\,T / 2\,\pi = c\,T^* / 2\,\pi = c / \omega$).

This circular structure appears static to an external observer because the vector '**H**' points always in the same direction relative to the tangent at every point of the circular path; this is because the Hyle angle of rotation is the same for each particular angular position of Hyle in the circle, when the structure is symmetric. That is how a matter ring, in perpetual rotation, appears as a solid and stable entity occupying a location in a 3D space.

Therefore, above structure explains coherently how nature builds rotating matter particles that appear static and independent from the passage of time or their spatial location; because of their circular symmetric configuration.

Any externally induced unbalance between 'T' and 'T*', caused by absorption, emission or collision, stresses directionally the circular structure and may also change its rotation plane; an external observer will see that the circle deforms and loses its static condition; the vectors start rotating in one direction or the opposite depending from which of the time variables is larger, 'T' or 'T*'.

Returning to equilibrium takes some time because the change of relative phase between the angular position in the circle and the angle of rotation must gradually return to zero and reorganize the structure; In other words, any discontinuity is an elastic phenomenon affecting the entire ring of Hyle, as a guitar string, when pulsed.

## The three angular velocities of a particle

The nomenclature shown in Figure VII will be used in this text.

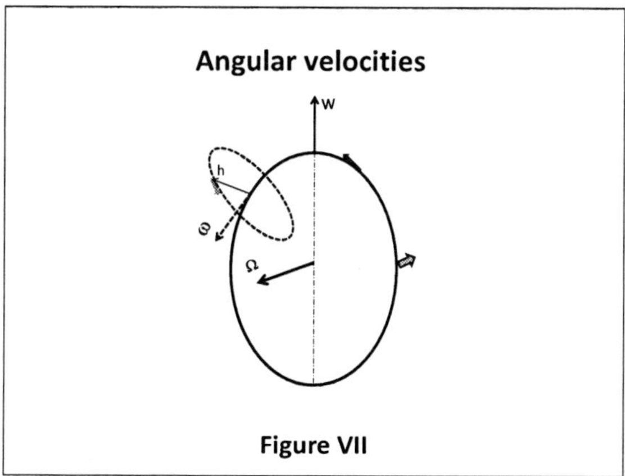


**Figure VII**

- ω is the angular frequency of the <u>Hyle vector</u> rotating on its plane.
- Ω is the angular frequency of the Hyle rotating on the perimeter from its ring.
- W is the angular frequency of the ring rotating on its diameter.

This figure shows clearly that the ring represents a gyroscope.

## The fundamental equation of matter particles

All matter rings are ruled by the basic natural "law of symmetry", which establishes that time elapses at the same rate in all dimensions, so vector 'H' turns $360°$ on its plane while Hyle completes circling the ring, whatever may be its plane of oscillation; in other words, the period 'T*' of a particle always equals the perimeter of its ring (T = T*); therefore, the radius of the ring in a symmetric particle is always (T* / $2\pi$), and its period is always 'T*'.

Colloquially we may say that all particles are rings, some may be heavier or larger, but their perimeter in travel-time is the same as their period of revolution in absolute-time; this assertion can be expressed mathematically by normalizing (symbolically) the vector function that describes the geometry of particles; therefore, the symbolic equation will be:

$$H / T^* = kte \qquad (4\text{-}9)$$

This will be known as the "law of similarity" which establishes that all matter particles "look as rings" with similar properties; their only difference is their individual radius; or its "scale"[31].

The magnitude of the constant "kte" is one [1/sec] in the natural system of units; but it may have other values depending on the system of units used.

By using this basic equation, the most basic laws in particle physics relating matter, energy, momentum, and other parameters; with mass, shape and motion, are found; just a consequence of this law, <u>with no previous knowledge about their existence.</u>

---

[31] It will be shown later that the "size" is inversely proportional to the "energy" contained in the ring.

## Chapter V

### Fundamental parameters of matter particles

This chapter shows how **the TH finds independently** the fundamental relations between most basic parameters in classical physics; as Einstein-Planck; mass-energy; energy and angular speed; spin and others; by just applying the laws of symmetry and similarity. This success validates the basic proposals and laws from the TH.

### Angular momentum

Classical mechanics defines the inertia of a mass that rotates at a distance 'R' from the center of rotation as ($I = m\,R^2$); the same formula defines the inertia of a rotating ring with radius 'R' and mass 'm'; so, the angular momentum '$L_R$' for a <u>ring rotating around its center</u> is:

$$L_R = I\,\Omega \qquad\qquad (5{-}1)$$

This equation may be read as ($L_R = I\,/\,2\,\pi\,T^*$) resembling the fundamental equation for particles (**H** / **T*** = kte) stating that *"the vector function 'H' per unit of time is the same for all particles in nature."* Let's assume that the magnitude of the natural parameter represented by "the Hyle vector function" 'H' is proportional to the Inertia 'I'; then, <u>any (symmetric) particle</u> will be ruled by the following relation:

$$I\,/\,T^* = I\,/\,T = kte \qquad\qquad (5\text{-}2)$$

Or: 
$$I\,\Omega = kte \qquad\qquad (5\text{-}3)$$

This fact establishes a fundamental equation for all material particles; this law establishes that: *"**the angular momentum for all matter particles is constant**"*[32]:

$$L_R = I \, \Omega = kte \qquad (5\text{-}4)$$

This classic formula represents the angular momentum of a rotating material ring of mass 'm' and inertia 'I' rotating around its center; this fact confirms that "H(t; τ)" is the particle's Inertia.

## The Planck constant

If we assume that the constant 'kte' is twice the reduced Planck constant ($h_r = h / 2\pi$), the above formula can be expressed as:

$$L_R = 2 \, h_r \qquad (5\text{-}5)$$

That is the classic formula known angular momentum of a material ring mass 'm' and inertia 'I', or a point mass rotating at an angular velocity '$\Omega$', as applied to a symmetric particle ($\omega = \Omega$).

The above formula shows that the modified Planck's constant represents what we will call the "radial" angular momentum of a particle; or in other words, it is a consequence of the "law of similarity" which then may be stated as:

*"**All material particles have a constant angular momentum equal to two times the modified Planck constant**."*

This relation also shows that the substance Hyle, when structured as a ring, acts as an inertial flywheel. Interestingly, this fact explains the

---

[32] This replacement does not mean postulating that the Hyle is mass; it only means that the magnitude of 'H' is proportional to a mass concentrated at the end of the radius of the circular path it describes; on the contrary, as it will be shown, this equation indirectly states that the mass from all stable particles is inversely proportional to the radius 'R' or to its perimeter 'T'.

origin for the implicit resistance of particles for changing their direction and position.

The equation (4-15) can be written as:

$$L_R = 2 \pi I / T = 2 h_r \qquad (5\text{-}6)$$

Or:
$$I / T = h_r / \pi \qquad (5\text{-}7)$$

## The axial momentum

In classical mechanics, the angular momentum of <u>rotation of the ring on its diameter</u> '$L_d$', is what we will denominate "axial" angular momentum:

$$L_d = I W / 2 \qquad (5\text{-}8)$$

When ($W = \Omega$), the relation between '$L_d$' and '$L_r$' is:

$$L_d = \tfrac{1}{2} L_R \qquad (5\text{-}9)$$

So:
$$L_d = h_r \qquad (5\text{-}10)$$

## The SPIN

If we define spin "s" as:

$$s = L_d / L_R \qquad (5\text{-}11)$$

Then, the spin for all matter particles is:

$$s = \tfrac{1}{2} \qquad (5\text{-}12)$$

This fact confirms that our proposal <u>assigning a ring shape to particles is correct</u>.

The parameter "spin" 's' is a mathematically generated generic reference term related to the directional attributes of a particle in quantum mechanics; remember that "s" is found by using diverse tools as "commutators" investigating the mechanical/geometric properties of the particle.

Direct experimental measurement of the radial angular momentum "$L_r$" is theoretically possible but difficult, while measuring the "axial" angular momentum "$L_d$" is performed by detecting the variation experienced by the plane of the ring as a response to controlled electromagnetic fields. This is the reason why experimental physics assumes that "$L_d$" is the "angular momentum" of a particle, which is understandable because the currently physics considers particles as spheres.

It is important remind that maximum theoretical (logically consistent) value for 'W' is ($\Omega / 2$), because one complete turn of the ring means returning to its initial state; hence there is "no rotation."

### The Einstein-Planck relation

The classical definition of energy of a rotating mass is:

$$\varepsilon = I\,\Omega^2 / 2 \qquad\qquad (5\text{-}13)$$

And:
$$I = L_R / \Omega = 2\,h_r / \Omega \qquad\qquad (5\text{-}14)$$

Then:
$$\varepsilon = h_r\,\Omega = h_r\,\omega \qquad\qquad (5\text{-}15)$$

***This is the well known Einstein-Planck relation.*** It is remarkable how easily was deduced by only applying the "law of similarity" to classical definitions.

This relation could be written representing much more genuinely the nature of particles, as:

$$\varepsilon\,T = h \qquad\qquad (5\text{-}16)$$

Were 'h' is the Planck constant.

## Mass

The concept of mass is "individual or complex particle structures that may occupy a spatial location"; consequently they may change it in a given time interval when an external influence is applied.

Considering that matter particles are rotating rings, inertia is the primary natural parameter from which mass concept derives. The classical equation defining the inertia of a ring is:

$$I = M\,R^2 \tag{5-17}$$

The particle radius 'R' is:

$$R = c\,T\,/\,2\,\pi = c\,/\,\Omega \tag{5-18}$$

Then:
$$I = M\,c^2\,/\,\Omega^2 \tag{5-19}$$

$$M = I\,\Omega^2\,/\,c^2 \tag{5-20}$$

Considering that the rings angular momentum is:

$$L_R = I\,\Omega = 2\,h_r \tag{5-21}$$

$$M = L_R\,\Omega\,/\,c^2 \tag{5-22}$$

$$M\,/\,2 = h_r\,\Omega\,/\,c^2 \tag{5-23}$$

Previously we have treated to the rotating Hyle as an inertial ring containing a mass 'm' distributed along the periphery and equivalent to half the mass of a conventional ring[33], then:

$$m = M\,/\,2 \tag{5-24}$$

---

[33] This conjecture is adopted to be coherent with Einstein's equation; but it demands an appropriate explanation. Possibly results from the translation of a rotating entity from a curved space to a flat space.

And this mass will be defined as:

$$m = h_r\, \Omega / c^2 \qquad\qquad (5\text{-}25)$$

Considering that in a symmetric particle ($\omega = \Omega$), then:

$$m = h_r\, \omega / c^2 \qquad\qquad (5\text{-}26)$$

Or:
$$m = h / T^* c^2 \qquad\qquad (5\text{-}27)$$

## Mass-energy relation

Considering that the mass of a particle is given by ($m = h_r\, \Omega / c^2$) in equation (5-25), we can replace terms it in the energy relation (5-16) just found ($\varepsilon = h_r\, \Omega$), and find that:

$$\varepsilon = m\, c^2 \qquad\qquad (5\text{-}28)$$

This relation represents <u>the famous Einstein equation.</u>

The previous analysis demonstrates that the proposals of the TH: that matter particles are *rings rotating at "light speed"*; is absolutely valid and reflects nature's reality.

## Total angular moment

The total angular momentum of a particle [34] is the vector addition of the radial angular momentum '$L_r$' and the axial angular momentum '$L_d$'; both are always perpendicular, then:

$$\mathbf{L_t} = \mathbf{L_r} + \mathbf{L_d} \qquad\qquad (5\text{-}29)$$

$$|L_t| = (|L_r|^2 + |L_d|^2)^{\tfrac{1}{2}} \qquad\qquad (5\text{-}30)$$

---

[34] Quantum physics defines spin 's' as an internal property of the particle related to '$h_r$'. It also defines a new variable 'S' and [$S2 = h_r2\,(s2 + s)$]. For ($s = 1/2$); ($S = 3^{1/2}\, h_r / 2$). Which differs in magnitude to the total momentum established by TH; relation between this two parameters is [$|L_t| = 2\,(5/3)^{\tfrac{1}{2}}\, S$].

$$|L_t| = h_r \left(|2|^2 + |1|^2\right)^{\frac{1}{2}} \qquad (5\text{-}31)$$

$$|L_t| = 5^{\frac{1}{2}} \, h_r \qquad (5\text{-}32)$$

## Particle's precession

In accordance to the previous analysis, all particles could be represented by a rotating flywheel, like a gyroscope reacting to externally applied torque, by changing the direction of its axis of rotation; this phenomenon is known in classical mechanics as precession. Consequently, complex structures such as atoms will have a particular angular momentum depending on their architecture.

## Photon momentum and mass

As the equation of Hyle constituting matter or signal is the same (H / T* = kte); one may feel tempted adopting the classical definition of momentum (p = h / c T) to photons, but we must remember that momentum implies mass, and mass implies "spatial residence." The photons are the particles that are creating the "space", never reside on it, so they cannot "change position." Moreover, a photon must become part of a particle, adding mass and energy to it; it must change, not be a photon anymore, and become part of the receiver mass.

Similarly, it is not possible apply the concept of spin to a one-dimensional linear entity travelling on a straight path; neither is appropriate assign nonexistent geometric concepts to a photon as: "rotation on its diameter"; mass; residence, and all other concepts which are implicit to spin definition.

## Elasticity

The concept of "elasticity" is originated by the Hyle in particles reacting to externally induced imbalance between its spatial length 'T' and the period 'T*'; this asymmetry between the two temporal dimensions generates an internal structural tension because every sector of the Hyle structure must obey the "law of symmetry", so the particle recovers the original equilibrium; in this case this fundamental and universal natural law establishes that all parts of a Hyle structure must move away one second of travel-time per elapsed second of absolute-time while maintaining its structural continuity instituted by the vector function '**H**'.

When an external influence compresses the Hyle structure, contiguous vectors of 'H' get closer, so internal travel-time 'T' differs from 'T*', creating a structural imbalance that generates the reaction of the structure increasing its "length" 'T' until it equals 'T*'; in photons this effect generates "gravity" and in particles expands their rings which breaks releasing energy.

This mechanics explains how recovering balance causes the particle release tensile energy residing in the travel-time domain '$\tau$' by decreasing its internal energy contained in the absolute-time domain 't'; explaining this in a simpler way, the ring breaks emitting a photon. The stress originated when a photon is absorbed by the ring, may be compensated by an externally applied opposite compression avoiding that the ring fractures, this is the basic mechanics governing atom "centricals"[35].

Now, after knowing these details, it makes sense the conjectures of De Broglie and Schrodinger describing particles as enclosures containing a vibrating substance; by adjusting the length they are able achieve resonance for a given energy content.

To better understand the phenomenon, let's suppose having an elastic material as gas confined in a tube; when the contained material undergoes a sudden compression it generates an elastic wave oscillates at the resonance frequency because the wave nodes

---

[35] The TH denominates "centricals" to the position occupied by electrons around the atomic nucleus; this is to differentiate from the description of "orbital" proposed by the QM, and the "orbits" described in Bohr atom.

coincide with the length of the enclosure; such is the case of a musical instrument like an organ. Any variation of the frequency of the exciting oscillation sound will momentarily shift the phase returning to the natural resonant frequency shortly after. Furthermore, if a ring perimeter varies, Hyle adopt a new wave resonance frequency imposed by the new length of the enclosure. When the media cannot support more compression, it releases some of its substance, returning to the initial oscillation period 'T*'.

In the case of a compressed photon, it has a rotation period '$T_o$*' that remains unchanged, so it expands adjusting its length until its length '$T_o$' in TT reaches the same value than '$T_o$*'.

Internally, the absorption of a photon by a particle generates a phase shift in the perimeter of the annular structure, which stresses the ring causing the structure not longer remain static, because its spatial phase does not match the temporary phase; a slow and continuous release of energy returns the particle to its initial symmetry[36].

As it will be analyzed later, all photons emitted are initially compressed due to the change of the subspace metrix at emission time; the photon expands for regaining symmetry, thus indirectly communicating the emitters distance to the receiver. This phenomenon, caused by the elastic property of the photon in observing the fundamental law of symmetry sets the radial metrix of "space", giving rise to what is conventionally knows as "the curvature of space" or "gravity."

---

[36] The generation of a node cancels a sector of vector function 'H' of the ring reducing 'T*') of the receiver; this imbalance is recovered by the emission of an incremental amount of energy send tangentially away.

## Hooke's law

To understand how a particle recovers symmetry, we must define "force" or "strength" as a property of matter describing the natural tendency of particles for returning to equilibrium in accordance to the fundamental law of symmetry.  This concept explains the origin of Hooke's law which states that small imbalances force is proportional to particles compression or expansion ($F = k\,\Delta x$):

$$F = k\,c\,\Delta T \qquad\qquad (5\text{-}33)$$

This is the way that macroscopically nature reacts to spatial deformation caused by external "forces." Once more; the most basic law of classical mechanics has its origin in the "symmetry law" of the TH.

## Natural system of units

Considering that we reside in a time based universe would be advisable develop a new scientific system of units, "**The Natural System of Units**", referencing to basic natural parameters, setting aside any artificial references historically established based in the subjectivity of our human perception. Current metric system base most definitions on the arbitrarily defined units of distance, and weight. A table containing the main parameters of physics is included at the end of this text.

This system of units only uses time as the basic variable; consequently various seemingly unrelated variables are measured by the same exponent of time units:

Since $[H] = [\,]$ (dimensionless), and the concept of 'H' is similar to the inertia 'I', then:

Inertia: $\qquad\qquad [I] = [\,] \qquad\qquad\qquad (5\text{-}34)$

Angular moment:  $[L] = [] / [sec]$                    (5-35)

Planck's constant:  $[h_r] = [] / [sec]$                (5-36)

The conversion factor 'c' is the ratio between TT and AT; and is one as ruled by the law of symmetry; then:

"Light speed" ($c = \tau / t = 1$)    $[c] = []$        (5-37)

Energy ($\varepsilon\, T = h$):    $[\varepsilon] = [] / [sec]^2$        (5-38)

Mass ($m = \varepsilon / c^2$):    $[m] = [] / [sec]^2$        (5-39)

Distance ($s = c\, T$):    $[s] = [sec]$        (5-40)

It will be also advisable define new precise relations involving AT or TT, as velocity, acceleration, vibration, etc. , and   other current natural parameters as: temperature, force, current, voltage, charge, etc…

Velocity ($v = c\, d\tau/dt$)    $[v] = []$        (5-41)

Acceleration ($a = dv/dt$)    $[a] = [] / [sec]$        (5-42)

Also define new values for constants and factors.

Gravitation constant ($G = g\, r^2 / M$)

$$[G] = [sec]^3 \qquad (5\text{-}43)$$

**<u>This page is intentionally left in blank</u>**

# Chapter VI

## Particle types

This chapter analyzes the shape, geometrical characteristics, and equations of the main particle types. Later, each one of the particle types will be assigned to their respective particles: neutrons, electrons, or protons.

Basically, there are two main branches of Hyle structures are:

a) The photons whose mission is carrying information from an emitter to a distant receiver, and

b) The matter particles which are the ones that build the complex material objects we observe in nature.

Both are described by the same equation, and their Hyle runs away from the emitting source increasing its "remoteness" at the same rate in accordance to the "law of symmetry"; furthermore, they define the geometry of a straight [37] line in their own "space", by the path they follow. A distant observer will perceive that photons paths are straight lines; and that matter particles are circles.

Previous chapters show that Hyle is an oscillating vector function '$H$ (t; $\tau$)' residing in a four dimensional time environment; its simplest

---

[37] The concept of straight line does not make any sense when there is not a "space." When a signal defines a distance when travelling apart from its source, the minimum distance trajectory is set by the signal path. When the particle is a photon, the path is a straight line; when the particle is matter, its path is a circle, and the circular sector becomes the shortest distance between two points on its perimeter.

structure is the signal carrier, the photon; while static ring shaped structures, are matter particles.

## Photon A and B structures

It is much more understandable to generically imagine the photon as a screw moving in the direction 'z' ($z = c\,\tau$) as it rotates at an angular velocity ($\omega = 2\,\pi\,/\,T^*$); conventionally we may say that they move at a speed ($c = |c|\;\mathbf{k}^\circ$) where $|c|$ is the magnitude and '$\mathbf{k}^\circ$' indicates the coordinate direction on which it moves; the wave number then is ($\kappa = 2\,\pi\,/\,\lambda = 2\,\pi\,/\,c\,T$). Depending on the plane of rotation we can define two types or structures, and describe them mathematically for ($T > \tau > 0$).

**Photon type "A" structure**. The plane of rotation is perpendicular to the length ('$\omega$' in the same direction of '$\mathbf{k}^\circ$'). This photon is circularly polarized.

$$H\,(z) = \sin\,(\kappa\,z)\;\mathbf{i}^\circ + \cos\,(\kappa\,z)\;\mathbf{j}^\circ \qquad (6\text{-}1)$$

**Photon type "B" structure.** The plane of rotation is longitudinal; ('$\omega$' is perpendicular to '$\mathbf{k}^\circ$'). This photon type is linearly polarized.

$$H\,(z) = \sin\,(\kappa\,z)\;\mathbf{i}^\circ + \cos\,(\kappa\,z)\;\mathbf{k}^\circ \qquad (6\text{-}2)$$

In both cases, the structure's travel-time increases 'T' on every period '$T^*$'; it has a wave length ($\lambda = c\,T$), and its angular velocity is:

$$\omega = (2\,\pi\,/\,T^*)\;\mathbf{k}^\circ \qquad (6\text{-}3)$$

$$H = 1 \text{ for } (0 < \tau < T)$$

Figure VIII shows photon types:

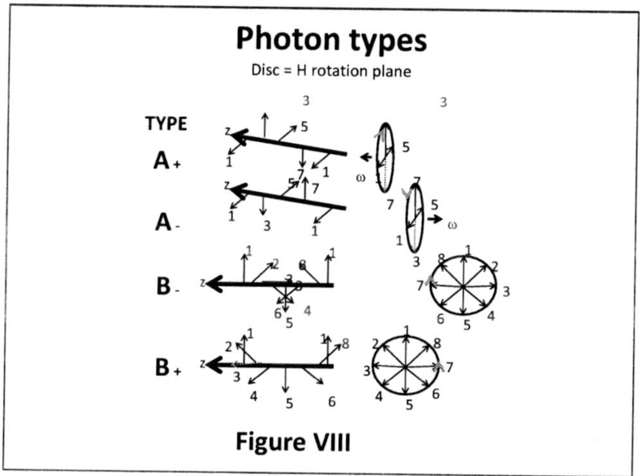


**Figure VIII**

Writing above equation in polar coordinates ($z = c\,\tau$):

$$\mathbf{H}\,(t;\tau) = r\,\mathbf{r}^{\circ} + \omega t\,\boldsymbol{\varphi}^{\circ} + (2\,\pi\,\tau\,/\,T)\,\mathbf{k}^{\circ} \qquad (6\text{-}4)$$

Radius 'r' is null, so:

$$\mathbf{H}\,(t;\tau) = \omega t\,\boldsymbol{\varphi}^{\circ} + (2\,\pi\,\tau\,/\,T)\,\mathbf{k}^{\circ} \qquad (6\text{-}5)$$

$$\mathbf{H}\,(t;\tau) = 2\,\pi\,[(t\,/\,T^{*})\,\boldsymbol{\varphi}^{\circ} + (\tau\,/\,T)\,\mathbf{k}^{\circ} \qquad (6\text{-}6)$$

'$H_{\varphi}$' is the AT angle of the period 'T*' of the photon and '$H_k$' is the TT position inside the structure, until it ends at TT 'T'.

Every photon ("A" or "B") carries charge and polarity information; future chapters will reveal which one carries emitter charge information; meanwhile the TH refers to them as photons type P (or Pp; for polarity photons) when they have polarity information; and as normal photons type N (or Np) when emitted by a particle with no charge.

Next chapters also explain the way photons communicate energy, and cause motion when absorbed by a particle.

The plane of rotation of the Hyle vector in a photon may not necessarily be perpendicular or parallel to its axis. Later it will be analyzed how a tilted plane may curve the trajectory of the photon until it reaches one of the stable structures ("A" or "B").

## Static structures, matter

Material objects are composed exclusively by three basic types of particles; protons, neutrons and electrons; from which only two of them appear being stable and have opposite charge, the proton and the electron. The neutrons, seemingly unstable, may get fractured generating one proton and one electron, suggesting that there were only neutrons at the beginning of time; this subject will be analyzed in full detail later. All other particles are unstable and have a very short life term. In later chapters, the TH proposes some hypothesis explaining why this may happen, and why the Hyle builds stable circular paths for these specific particles.

The circular path of Hyle form rings with a perimeter equal to the wavelength (c T). Symmetrical particles build "static structures", with all their vectors always pointing in the same direction relative to the tangent of the ring and have the same magnitude (equal to one) all over the perimeter; they are not standing waves which change amplitude with time. Geometrically, when (T = T*), the phase of the vector on the helix coincides with the angular position, in this way a remote observer capable "seeing" vector 'H', simply will perceive a static ring with a given radius and constant thickness that represent the unit amplitude of vector 'H'; he will not even notice that 'H' is circling at the speed of light[38].

---

[38] We use improperly the term "speed of light" once more in order to avoid complicating the explanation. This rotation establishes the independence of this entity, being the origin of the apparent "solidness" of matter.

In this way, nature builds stable space-time structures that remain unchanged indefinitely until interacting with another particle. Consequently the Hyle ring is perpetual; or at least extremely stable. When ring diameter changes, it causes a structural stress that destroys the balance between revolution AT and TT perimeter, so the structure immediately ceases being static.

This balance could be influenced or restored by:

- By applying an external force that will restore its perimeter; one way for compensating centripetal stress is rotating the ring around its diameter, so that the centrifugal force expands the perimeter; another opposite way for restoring equilibrium is to compress the ring by electric or mechanical means.

- By decreasing or increasing the internal energy when emitting or absorbing the energy ($\varepsilon = h / T^*$) from a photon respectively. Absorption decreases the period 'T*' in relation to the perimeter 'T' (wavelength), meaning that the diameter becomes too large, the generated stress in the perimeter of the ring, breaks it.

**<u>Angular speeds nomenclature for rings</u>**

The Hyle constituting a material particle is described as a rotating vector function 'H' on a plane with an angular speed ($\omega = 2\pi / T^*$), while moving tangentially on the direction of 'z' and describing a circular path. The Hyle completes circling the ring in a TT equal to 'T'; therefore it turns around at an angular velocity ($\Omega = 2\pi / T$). Moreover, the ring can rotate about its diameter at an angular velocity 'W'. The figure below describes this nomenclature, ruled by the right hand convention.

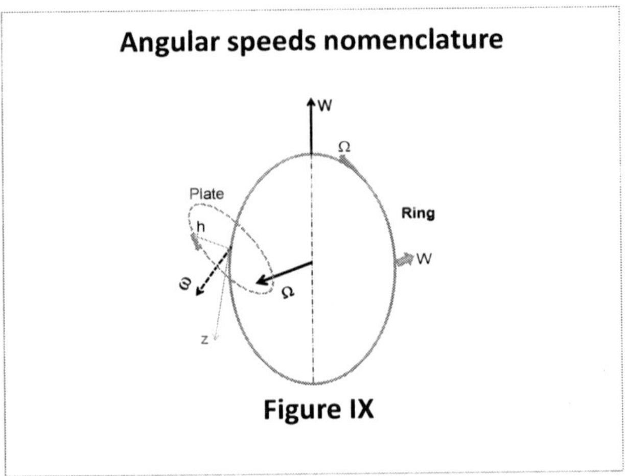



**Figure IX**

Symmetrical particles look static for an outside observer as the angle of vector '**H**' always coincide with the angular position on the perimeter because (T = T*); whereby, the direction of vector '**H**' depends only on the initial angular position and the angle between 'ω' and 'z'.

## Nomenclature of particle types

Coincidentally, the TH finds that there are three possible ring structure types, and there are three particles in nature: the proton, electron and neutron. All other particle types are unstable or antimatter. When the Hyle's rotation plane is neither parallel, nor perpendicular to ring's plane, the particles are unstable; there is a natural reason for these structures break apart, which will be explained later. The type of structure that corresponds to each particle will also be defined in the respective chapter.

The angle between ring's plane and Hyle's rotation plane defines the particular structural shape of a particle type; the following figure illustrates the three types of particles when the tangential angle is $0°$ or $90°$; it illustrates the shape by showing the position of Hyle vectors on the periphery of the ring.

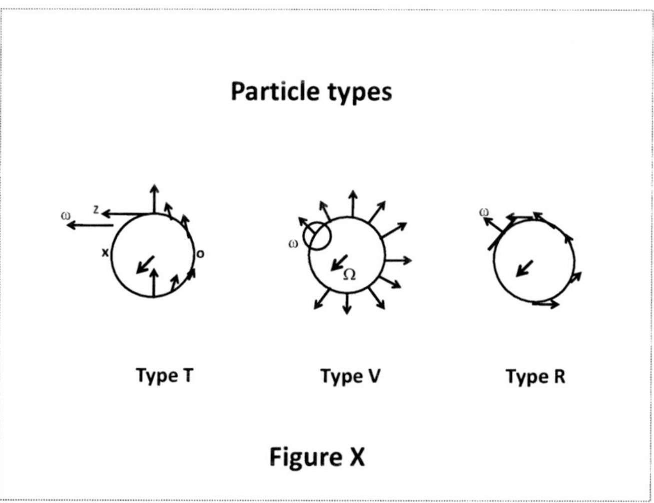



**Figure X**

A tangential type particle, or **type "T"**; the vector '$\omega$' is tangent to the ring. When it points to the same direction than 'z' is "T(+)", and when points to the opposite direction is "T(-)"; the Hyle exhibits a radial "plane of rotation" (or disc), perpendicular to the ring's plane.

A vertical particle type, or **type "V"**; the vector '$\omega$' is perpendicular to the plane of the ring (vertical). When it points to the same direction than '$\Omega$', is "V(+)." The disc plane coincides with the plane or the ring.

A radial particle type, or **type "R"**; the vector '$\omega$' points in the same direction of the radius of the circle; and is "R(+)" when points to the center. The disc plane is perpendicular to the radio, i.e. tangential.

As shown, all three structures have two alternative versions: positive when '$\omega$' has the same direction than that of reference vector and negative when opposite; although, this characteristic does not define the charged particle sign; it is only useful analyze their geometrical symmetry.

### Ring's general equation

The general particle's equation is;

$$H\,(t;\,\tau) = (\sin \beta\ i^{\circ} + \cos \beta \cos \gamma\ r^{\circ} + \cos \beta \sin \gamma\ z^{\circ}) \qquad (6\text{-}7)$$

$$\beta = 2\,\pi\,\omega_o$$

$$\gamma = \text{arc cos}\,(z^{\circ}{}_{dot}\,\omega^{\circ})$$

$$0 < \gamma < \pi/2$$

$$|H| = 1$$

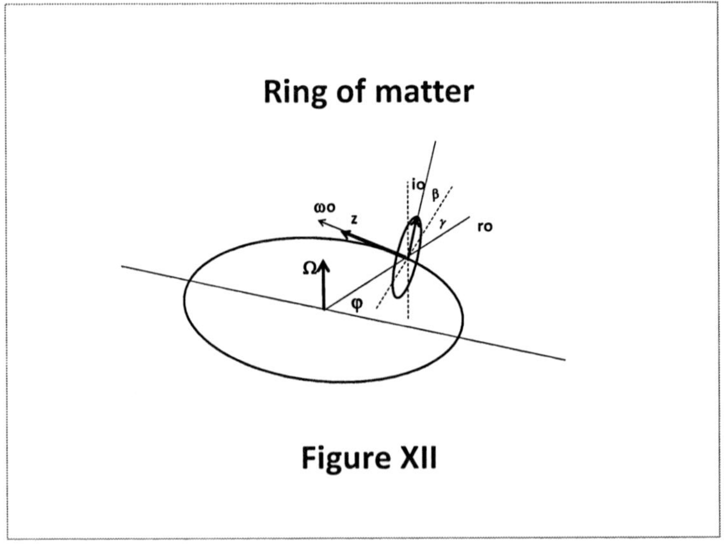


**Figure XII**

As shown in **Figure XII**; '$\beta$' is the angle between '**H**' and the ring plane, as it rotates with an angular frequency '$\omega$'; '$\varphi$' is the angle between the plane of rotation with the ring's initial position. The structure completes a complete rotation in time '**T**', at an angular speed '$\Omega$', where ($T = 2\,\pi\,/\,\Omega$)

The angle '$\gamma$' defines the type of structure; it is constant unless the ring is subject to stress. In any stable particle ($\omega = \Omega$) as ($\beta = \varphi$), which means that vector '**H**' always points in the same direction at the same angular position on the ring, instituting matter as a "time independent structure."

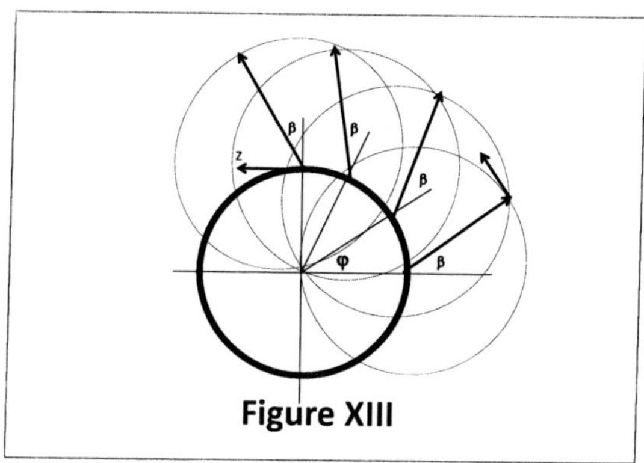


**Figure XIII**

As a way of illustration, **Figure XIII** shows the simplest equation of 'H' for a type V particle, where the disc plane coincides with the ring plane:

The angle '$\alpha$ between' **H** and the radius is ($\alpha = \beta - \varphi$):

$$\mathbf{H} = \sin \alpha \; \mathbf{x}^\circ + \cos \alpha \; \mathbf{y}^\circ \tag{6-8}$$

$$\beta = \beta_o + \omega_o \tau \tag{6-9}$$

$$\varphi = \Omega \tau \tag{6-10}$$

$$\alpha = \beta_o + (\omega_o - \Omega) \tau \tag{6-11}$$

As ($\omega_o = \Omega$); $\qquad \alpha = \beta_o \tag{6-12}$

**Figure XIII** shows a type V structure because of the simplicity of drawing it on a plane:

$$\mathbf{H} = \sin[\beta_o + (\omega_o - \Omega) \tau] \; \mathbf{x}^\circ + \cos [\beta_o + (\omega_o - \Omega) \tau] \; \mathbf{y}^\circ \tag{6-13}$$

When the structure has ($\omega_o = \Omega$), it produces a static configuration for **'H'** in the ring.

$$H = \sin \beta_o \, x^\circ + \cos \beta_o \, y^\circ \qquad (6\text{-}13)$$

The particle's ring may rotate <u>on its diameter</u> with an angular velocity **'W'**, always perpendicular to '$\Omega$'.

**Particle type T (+)**; where ($\gamma = 0$), shown in **Figure XIV**

In this structure '$\omega$' is parallel to '**z**'; charge may be defined as positive (+) when both have the same sense of rotation, and negative when opposite. In this structure '$\omega$' is perpendicular to '$\Omega$'.

Applying to the equation (6-7), the value of ($\gamma = 0$) and a static structure condition ($\omega = \Omega$):

$$H = \cos \varphi \, r^\circ + \sin \varphi \, i^\circ \qquad (6\text{-}14)$$

This is also a time independent geometric structure.

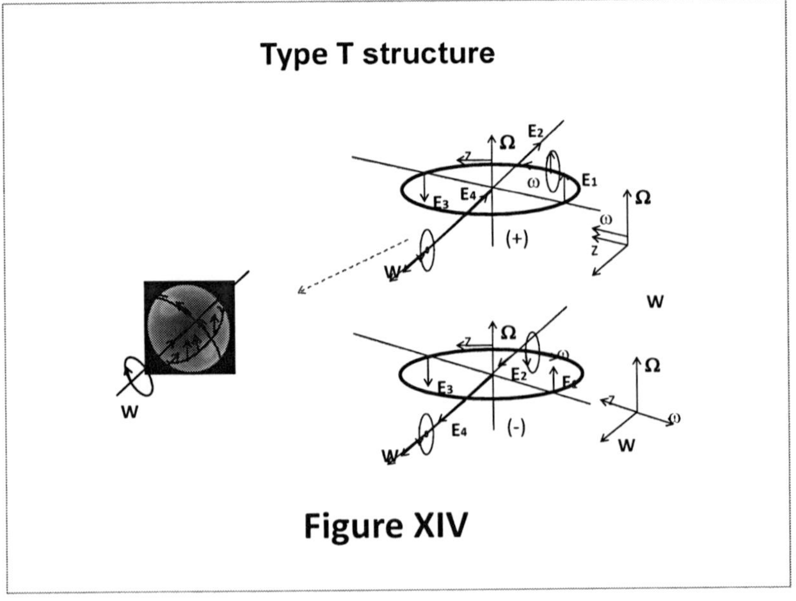


**Figure XIV**

**Particle type T(-)** with $(\omega_o = - z_o)$ $(\gamma = \pi / 2)$; lower part of **Figure X**.

As, $(\omega_o = - \omega_o)$ in equation (6-14), then:

$$H = - \sin \varphi \; i^o + \cos \varphi \; z^o \qquad (6\text{-}15)$$

By rotating the ring $180^o$, $(\varphi = \varphi_o + \pi)$ we get the same equation as T(+); thus, they are symmetric structures; it is enough to rotate one of the rings on the two axes (vertical and horizontal) and obtain the same structure.

**Type R of structure**.- where $(\gamma = \pi / 2)$, is shown in **Figure XV**.

Replacing in Eq (6-7): $\quad H\,(t;\, \tau) = \sin \varphi \; i^o + \cos \varphi \; z^o \qquad (6\text{-}16)$

Here, '$\omega$' is radial and perpendicular to '$\Omega$' and '$z$'.

For R(-),$(\omega_o = - \Omega)$: $\quad H = - \sin \varphi \; \varphi^o + \cos \varphi \; k^o \qquad (6\text{-}17)$

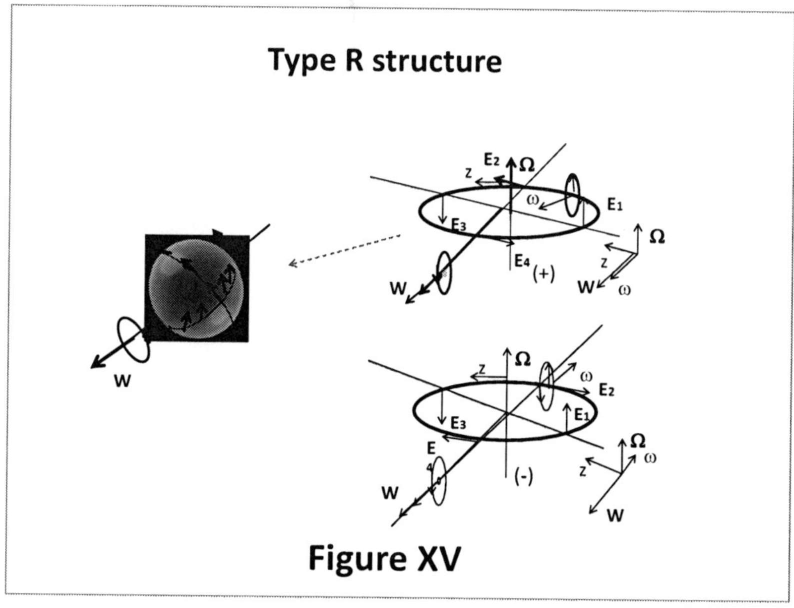


Figure XV

This pair of structures is also symmetric as in the previous case.

**Particle type V** shown in **Figure XVI**

This structure has '$\omega_o$' parallel to '$\Omega$' (or negative in a second case), both perpendicular to '**z**' and '**W**'.

Applying the general equation (6-7):

$$\mathbf{H}\,(t;\,\tau) = \sin\beta_o\, \mathbf{z}^\circ + \cos\beta_o\, \mathbf{r}^\circ \qquad\qquad (6\text{-}18)$$

The geometry of the structure depends on the value of '$\beta$', the most interesting angle values are ($\beta_o = 0$ or $\beta_o = \pi$) for structure R(+):

$$\mathbf{H}\,(t;\,\tau) = \pm\, \mathbf{r}^\circ \qquad\qquad (6\text{-}19)$$

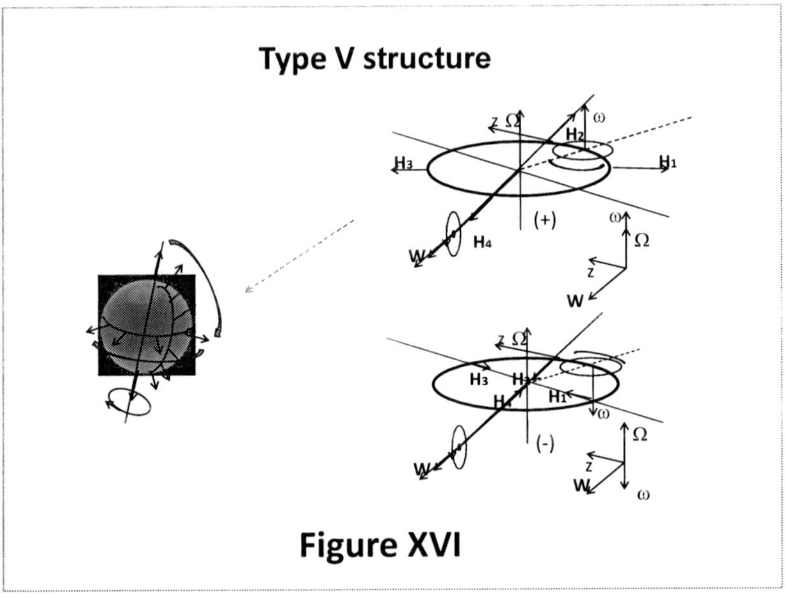



For reverse rotation ($\omega_o = -\,\Omega$):

In this case ($\beta = -\,\omega_o\, t$) and ($\varphi = \Omega\, t$):

$$\mathbf{H}\,(h;\,t;\,\tau) = \sin(\beta - \varphi)\, \mathbf{z}^\circ + \cos(\beta - \varphi)\, \mathbf{r}^\circ \qquad\qquad (6\text{-}20)$$

$$\mathbf{H}\,(h;\,t;\,\tau) = \sin(-\,2\,\varphi)\, \mathbf{z}^\circ + \cos(-\,2\,\varphi)\, \mathbf{r}^\circ \qquad\qquad (6\text{-}21)$$

$$\mathbf{H} = -\sin(2\,\varphi)\, \mathbf{z}^\circ + \cos(2\,\varphi)\, \mathbf{r}^\circ \qquad\qquad (6\text{-}22)$$

Which is symmetric with R (- π/2)

**Figure XVII** shows a comparative illustration for all configurations.

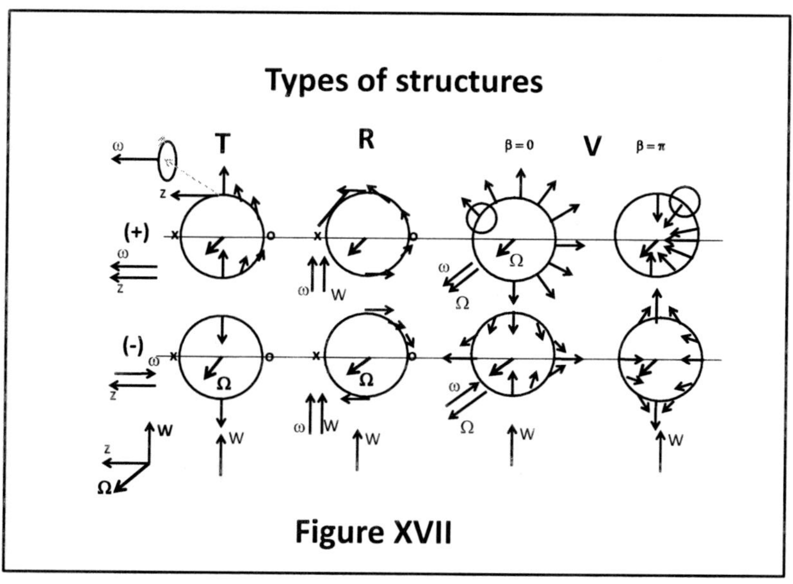



**Figure XVII**

In accordance to this proposal, protons, neutrons and electrons "look" significantly different.

There may also be hybrid particle with the Hyle rotation plane not perpendicular neither coplanar with the plane of the ring that could form different structures, but as it will be analyzed later, there are only three stable particles in nature, which makes us presume that non-orthogonal planes migrate to a stable structure.

This page is intentionally left in blank

## Chapter VII

## Basic concepts on emission and reception

Previous chapters explained how the term "space" defines a contingent entity generated by the presence of matter inside a volume containing distant emitters and receivers continuously interchanging signals. Signals are information carriers that communicate the receiving particle some basic properties from the emitter by influencing appropriate changes in the Hyle structure of the receiver; for instance, attracting it or repelling it in accordance with emitter mass and its charge.

Both, the emission and reception is always performed by individual particles that eventually are part of a complex object. In any case the emission is the result of photon absorption producing imbalance between the period 'T*' of the Hyle (in AT) in the particle and its "length" 'T' (in TT). The internal stress generated returns to equilibrium by releasing or absorbing energy.

Under this perspective space is really a field that could be described as: "a volume whose potential properties at every location are established by the low energy photons emitted by distant particles, affecting a receiver on that location." The space that surrounds an emitter exists as far as it emits signals; moreover, when there is no receiver capable of absorbing these signals, the concept of space and its dimensions fades away. A flat space, with no matter or signals cannot exist; it is just a useful entelechy that facilitates mathematical analysis by locating geometrically particles in accordance to our human perspective; the only place in nature that resembles the concept of "flat space" is the intergalactic space where there are just a few photons crossing it from all directions building a delicate entity.

Emitters do not receive back their own signals; they just send explorers that will eventually arrive to a receiver able send myriads of signals in all directions, so some of them may be reflected or may originate the emission of a photon that will return to the emitter, but the emitter will never know which of them it is. To define "space" as a stable natural entity there must be a permanent interactive communication always detecting not recognizable signals.

The term "composite-space" sometimes will be used, referencing an environment subject to the influence of several sub-spaces. The properties of every point in the composite-space are statistically defined by the <u>weighted qualities, quantities, and direction</u> of photons actually passing through that location.

The natural signal emission from large objects is the consequence of a permanent internal signal interchange between its atoms; consequently, the characteristics at a given point in a composite-space depend on the temperature of the emitters. Additionally emitted signals must be non destructive; space exists only in the vicinity of an object whose spectral emission is not much wider than solar spectrum band-width. A gamma ray source (or powerful X rays emitter) may destroy the receivers; consequently it cannot constitute a place where an interchange of signals may generate a "space" around it; it will be just the TH version of a "black space" inside the intergalactic space. Massive matter concentrations generating a stable gravitational sub-space are mainly represented in nature by stars surrounded by their planetary systems; consequently, the meaning of "space" is closely related to gravitational "sub-spaces" produced by bright objects at high temperature; a low temperature massive object will not have as much influence on the surrounding environment as compared with a similar bright one. The attraction applied by a massive black object on a near object will diminish as the quantity of emitted photons lowers, because its relative influence is weaker.

When two Hyle structures meet, the meeting point involves two internally defined rectilinear trajectories. When the energy content of one of them is much larger than that of the other, the first one sets the spatial geodesics. When two particles have a similar energy, they will collide elastically as billiard balls; moreover, when two similar particles collide, both can be destroyed when meeting al high relative speed. A material ring can absorb a low energy photon, or be destroyed by a high-energy photon. Furthermore, there is an additional possibility that will be analyzed later describing similar particles sharing a common empty space; a node.

The non destructive photons energy is tenths of thousands smaller ($10^{-4}$) than the energy of an electron or even smaller than a proton. Considering that <u>most natural signal emissions are produced by electrons in atoms</u>, geometrically, the emitter can always be considered a dimensionless point in space. On the other side, a photon is also a geometrical line with no thickness, this is how signals travelling through "space" may enter to the "own-space" from a particle. The arriving photon may touch the particle ring and be wrapped around the particle ring[39].

There are occasions when a particle may emit a lower energy particle; for instance, when a neutron divides itself, it emits an electron in the well known process of $\beta$ decay. It is possible that in this case the emitted high energy photon wraps on itself because the plane of the emitted Hyle is initially a curved path of the emitted photon which finally closes on itself forming an electron. It is important to mention that the particle emitted by this process involves three "energy dimensions"; <u>the internal energy</u> defined by its period 'T*' in the AT

---

[39] The absorption process should be studied in deep by analyzing the interactive phenomenon between the two Hyle vector functions '**H**', one of which is so long that it is coiled on the ring until it concentrates enough energy to build a node.

dimension, the <u>internal stress</u> in the TT dimension, and  the <u>kinetic energy</u> ($m\,v^2$) of a mass travelling through space at a speed '$v$'[40].

The following emission/reception analysis will be restricted to low energy photons able to institute "space."

## The nodes

The interaction between signals and particles is determined by the simple mechanics of producing empty spaces within the particle structure, annulling vectors along the structure; these empty spaces will be denominated "nodes."

The Hyle of any particle is a substance travelling always on a straight path, although perceived as a circle in the case of the matter particles. The Hyle is a continuous substance mathematically described as a rotating vector function pointing to different direction while advancing; the amplitude of the vector is binary, with a value equal to one within particle space-time limits, so the Hyle vector '**H**' from particle rings and photons have the same magnitude equal to "1" ($|H| = 1$). The energy contained by a particle is defined by the angular velocity. Under this perspective we could describe any particle as a continuous series of "energy packets" as vectors pointing to the direction of '**H**' in that point, and containing an arbitrarily preset amount of energy[41].

When two particles share the same space point, their equivalent energy packets meet so their vectors will add, and when their summation result is zero these segments will cease existing, generating an empty space along the ring while the section from the arriving photon disappears.

---

[40] This point of view should be taken into account when investigating the existence of neutrinos defined contradictorily as particles that are *"very small and with low energy"* ($T^* \ll T$?)

[41] This model is not strictly valid, but may be useful to visualize the phenomenon.

Figure XX describes the interaction in the point where photon vector "5" opposes to the ring vector "A"[42]. The concept of nodes will be analyzed in full detail in next chapters; basically a node is the empty space within the ring periphery, generated by canceling a section of the ring structure, so simultaneously producing an instantaneous decrease in the period 'T*' of the Hyle in the receiver.

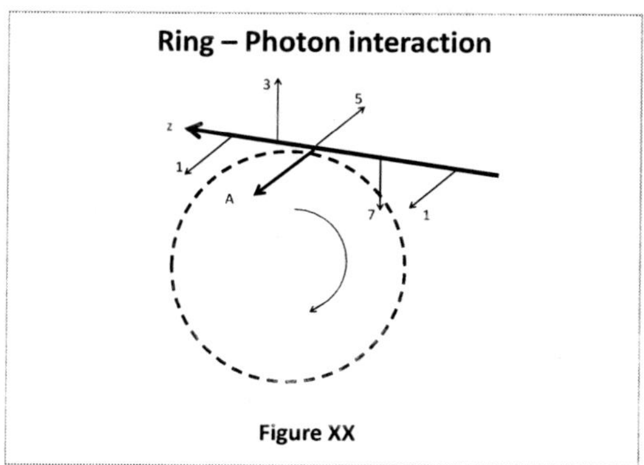


Figure XX

There are two effects produced by the emergence of a node within the particle ring:

a) It reduces instantaneously its period 'T*', accelerating its rotation and increasing its internal energy, and

b) It creates an empty section on the ring, stressing it because its radius cannot reduce instantaneously because it takes time to the internal elasticity in order to recover symmetry by compressing the ring.

---

[42] When two Hyle vectors share a spatial position, they will add each other. When the addition result is larger than zero (could be larger than 1 or a fraction) two Hyle structures do not mix and remain independent, so the particles collide. In the case of photon absorption, the Hyle of the photon is wrapped on the ring as an independent entity travelling on a parallel path accumulating energy until raising enough energy to cancel an opposed and equivalent energy sector from the ring; when this happens, the photon section disappears, generating a node on the ring.

This explanation describes how a particle increases its energy by absorbing a photon which ends up disappearing in the process. It is interesting observe that the particle rotates faster and becomes smaller when absorbing Hyle; this behavior contradicts our intuition; we are accustomed imagining that objects become larger when increasing their mass.

The described process repeats until all the photon energy is absorbed, distributing nodes along the periphery of the ring.

Summarizing the absorption phenomenon, the reception mechanics decreases the period 'T*' of the particle while the perimeter length remains, so generating an internal asymmetry manifesting as a centrifugal force which breaks the ring. Considering that the elastic perimeter reduction process, by the action of its internal elasticity, takes too long for recovering equilibrium; the centrifugal force becomes more important, liberating energy in the form of a photon, expelled tangentially.

The emission process rebuilds the nodes by generating energy segment leaving the ring tangentially while recomposing the missing energy vector within the particle[43]. Consequently, the Hyle leaking from the particle will rotate on the same plane than the disc of the particle at that point of its ring, so a particle whose Hyle changes its angular emission point during the process, will generate a new particle because the emitted structure will separate from the structure following a curved path; this behavior will be analyzed later, explaining the mechanics of the beta emission.

Additionally, as it will be explained later in detail, absorbing an asymmetric photon will generate an asymmetric distribution of nodes

---

[43] Obviously, the rotation plane direction of the emitted vector function will be defined by the much more energetic function of the particle in that position.

in the ring, deforming (and stressing) a section of the ring more than others.

## The absorption and emission process in free particles

A symmetric particle ($T = T^*$) absorbing a photon decreases its rotation period '$T^*$' (increasing its angular speed '$\omega$') to a new value '$T_1^*$' that increases its internal energy because ($T_1^* < T^*$), while applying an expansive force to the ring that does not change its size immediately because the empty spaces generated by nodes in the ring substance take some time  getting closed; its internal elasticity initiates a slow process that compresses it until the ring recovers symmetry, so the particle remains asymmetric ($T > T^*$) for many revolutions. It has two alternatives for recovering stability:

a) Increase '$T^*$' by emitting any excess of energy.

b) By reducing its perimeter '$T$' by the action of its internal elasticity.

Consequently, any free particle cannot remain asymmetric for a long time; it regains its symmetry before breaking, by emitting a photon. While its perimeter decreases slowly the rotation period '$T_1^*$' increases until reaching the original equilibrium value '$T^*$'.

The emitted photon increases its distance at the rate of one ring perimeter per revolution; the fast rotation at period '$T_1^*$' decreases until reaching the original value '$T^*$'. The phenomenon resembles a thin thread which unwinds from a rotating spool [44] decreasing its angular speed; the length of unwind thread per unit of time will become also smaller. Considering that the emitter normally stores tens of thousands times energy than emitted photons, its radius change is negligible during the whole emission process.

---

[44] The tangential expulsion of an "energy package" which is part of a continuous substance implies increasing spatial spectral width, while building a continuous entity without cuts. This process is so sensitive that deserves deeper investigation.

## Signal emission

When evaluating this section, the reader should not try to instinctively build a mental image of a space and geometrically locate emitters and receivers on it; this is in order to avoid developing erroneous concepts. When there is not a continuous communication, all particles in nature are isolated and static, and all signals already emitted increase their TT as AT passes, which means that the signals permanently travel away from a steady reference point always coinciding with a static emitter at the moment of emission, but not necessarily crossing a preexisting "space."  All influences on the receiver as forces, and spatial properties as "distance", are established by the signal as <u>potential informative parameters</u> that will cause a change only when absorbed by a receiver.

For our spatial-oriented mind it is difficult to explain phenomena without referring to locations and positions. As it was mentioned before, in this study proper semantics sometimes will be sacrificed for improving clarity of the explanation. Using the appropriate semantics we should say that photons occupy the three "travel-time" dimensions, expanding this potential entity continuously as time passes; we can express the same idea by using conventional language, saying that all emitted signals are <u>enclosed within a sphere whose radius grows at speed 'c'</u>.

Any signal sent by an object travelling away from the observer at a speed higher than 'c', will never arrive to it as explained in the following paragraph; this is why for analysis purposes, the potential "space" surrounding an emitter can be considered as its own sub-space; as a part of its individual structure.

Any particle is always at rest while all other objects in the universe change position around it. When a particle emits a photon, it departs at a speed "c." By assuming that a particle is part of an object with many particles continuously stimulated, the particle will keep emitting new photons travelling away at the same speed, producing a continuous sequence of photons; if the object emits photons periodically in all directions, a new spherical surface centered at the emitter is created with every new group of photons released; the radius of the new sphere will be concentric and also grows at a speed 'c'.

From the observer's perspective, when a emitter moves away at a speed 'w' greater than 'c'[45], will never be detected because the surface of the emitted field is not growing fast enough; effectively, the photons <u>travel away</u> from the observer at a speed (w − c), never reaching it. This phenomenon, properly stated will be described by saying that the signal TT increment is smaller than the same of the receiver.

When the relative distance to the emitter keeps increasing, but it slows down (w < c); the spherical surfaces of emitted photons will remain being concentric to it because the emitter is always static. From the observer's perspective the emitter subspace is approaching at a speed $(c - w)^{46}$.

The most important photon task is that it establishes a spatial reference at emission point. This characteristic, consequence of the "law of symmetry", provides to the signal carrier the capability to <u>define</u> (although, not communicate) the travel-time when reaching a receiver. Considering that the photon leaving the emitter increments its "distance" from the emission point as absolute-time passes; the

---

[45] This statement contradicts to the TR, under the TH principles, any self propelled object is isolated from everything, so it can increase its relative speed indefinitely since there is no need to "bend an already existing space." An observer will perceive its signals shifting to infrared and finally disappearing.

[46] When the emitter decreases it speed, relative to the receiver; then, the receiver enters to the outmost spherical surface from emitter's sub-space intercepting the photon layers that did not arrive before. Therefore, the observer will perceive an increasing bright light.

emitter may disappear; but even in that case, its sub-space will remain being concentric to the point where emitter was located at emission time; the only noticeable effect is that after the emitter disappearance, not new photons will be emitted.

The released energy defines the period of a signal. As analyzed before, a free particle absorbing a photon will release a photon with different structure than the received one, but having the same energy[47]. Considering that most of naturally emitted signals are originated by atomic electrons changing orbital energies, the period of their signals is closely related to the specific atom's internal architecture, and not always to the absorbed energy.

Distance information depends on the signal TT between emission and reception; this information is independent of the signal's period. All sub-spaces are independent of the emitted photons energy; obviously a high energy photon will need to complete more revolutions to travel the same distance (TT) than a weak photon.

It is important to remember that the emitter is always a ring containing much more energy than the tangentially emitted photon; consequently the emission process takes a large number of ring revolutions.

## Signal Reception

The most important concept in signal reception is that the absorption process is a local event between a particular photon and an individual particle; it is independent from travelled-time or the relative speed of the emitter; is the information the one that modifies some properties of the receiver. Expressing it in conventional language we may say that absorption is never a relativistic phenomenon.

---

[47] It explains how free electrons can communicate their electric properties.

When an emitter sends <u>two independent signals</u> in a time interval 'dt', a distant moving receiver will compute the time between events 'd$\xi$' distorted by the vector addition of the change in TT 'd$\tau$'; as given by the relation:

$$d\zeta = (dt^2 + d\tau^2)^{1/2} \qquad (7\text{-}1)$$

Within an imaginary flat space environment the distance between a stationary emitter and a static receiver does not change (d$\tau$ =0), and (d$\zeta$ = dt). By applying this concept to the period detection of an arriving photon as an event and the arrival of its tail after a period 'T' as a second event, when 'T' does not change, it will be read by a receiver also with a duration equal to 'T', independently from the travelled time.

It is very important to comprehend that the emission of the point and the tail of the photon <u>are not two isolated events</u> because the photon is a continuous structure whose length is established by natural rules [48] set by the emitted energy and the ring structural geometry; the description of reception process does not refer to two independent events, but to <u>the gradual absorption of a discrete entity</u> of certain size. This basic natural phenomenon confers to signals the capability to be an accurate information carrier, because is an independent structure whose architecture only contain the characteristics given to it by the emitter at the moment of emission.

A photon does not have the capability to store the elapsed TT, but it is a Hyle structure able to communicate faithfully its length to the receiver during the absorption process; this information is used to decode relative speed, which may be linked to distance. In this way,

---

[48] Two independent events will not transmit faithfully original time intervals since its detection will have to account for the vector quality of travelled time increments between the emitter and the receiver, caused by their relative speed.

nature builds around emitters "sub-spaces" not as "distances fields" but as "velocity fields."

## Relative position and velocity detection

In order to locate the position of an object the receiver needs a signal that communicates directly or indirectly the direction and distance of the source, specifying the three-dimensional spatial location of the object. When analyzing an individual interaction, we can simply use the line of sight as the unique reference coordinate; then the distance will be provided by the signal TT '$\tau$' which may change by '$\Delta\tau$' when relative position changes after an absolute time interval '$\Delta t$'; its ratio will define relative velocity as ($v = \Delta\tau \,/\, \Delta t$). It is important to remember that the particle ring has the geometric capability to detect relative direction of arrival.

The signal TT '$\Delta\tau$' separating two particles could be converted to conventional metric units of spatial distance by applying the conversion factor 'c' to the travel time ($c\,\Delta\tau$); the ratio between TT and AT is conventionally known as the "speed of light" equal to ($c\,\Delta\tau \,/\, \Delta t$); its value in the natural system of units will be one for a symmetric photon. The same conversion applied to travel-time '$\Delta\tau$' will find ($\Delta s = c\,\Delta\tau$), then the relative velocity will coincide with classical definition ($v = \Delta s \,/\, \Delta t$)[49].

It is obvious that the only way for nature to detect the emitter relative speed is by determining the "length" '$T_o$' of the absorbed photon; the way to do this is by measuring the absorption process duration '$\Delta t_1$' with the internal clock of the receiver. When the photon is symmetric ($T_0 = T_o{*}$) and the receiver does not move with respect to the emitter, '$\Delta t_1$' will equal the length of the photon ($T_0 = \Delta t_1$).

---

[49] It is important to note that the minimum distinguishable '$\Delta\tau$' is the photon length '$T_o$'.

When a transmitter approaches to the receiver, the absorption process takes a shorter time, because the receiver decreases its distance to the emitter by '$\Delta\tau$' while the photon increases its travel-time by adding one photon length '$T_o$'.

The important fact in the absorption phenomenon is that the decrement/increment of detected TT ($\Delta\tau$) during the absorption time ($T_o$) will communicate the difference of the actual signal period '$T_o$' and the nominal length that should be equal to '$T_o{}^*$' in a symmetric photon; then ($\Delta\tau = \Delta T = T_o{}^* - T_o$); therefore, the ratio of distances ($\Delta\tau / T_o = \Delta T / T_o$) during the interaction time '$T_o{}^*$' represents the ratio of relative speeds with respect to the same reference point, then:

$$\Delta T / T = (\Delta T / T_o{}^*) / (T / T_o{}^*) \qquad (7\text{-}2)$$

Or: $$\Delta T / T = v / c \qquad (7\text{-}3)$$

This formula is the well known Doppler's relation; the conceptual differences between Doppler and TH perspectives will be analyzed in the next paragraph.

## Doppler's effect

Exploring the conventional explanation of how an observer perceive signal's frequency shift when the emitter is moving, allows him to detect relative speed. The importance of this phenomenon demands a profound analysis.

Imagine a photon as a wave with a period '$T_o{}^*$'; the receiver will read this constant period also as '$T_o{}^*$' when the emitter does not move.

When the relative distance emitter–receiver changes at a given relative rate ($\Delta T / T$), the receiver will read the signal's period as being ($T_o = T_o{}^* \pm \Delta T$). The interaction between particles is local; consequently, is not subject to relativistic time-distortion corrections.

Explaining it in conventional language, the speed at which the photon interacts with the receiver will be $(c \pm v)$, 'v' being the emitter-receiver relative speed at the time of emission.

Considering that <u>the receiver is always stationary</u>; is the emitter that moves at a relative speed 'v' carrying with it all its own "sub-space"; as photons fly away from the emitter surface at a speed 'c', the relative velocity of the photon wavelength $(c\, T_o)$ intercepting the receiver is equal to $(c + v)$, depending if the emitter gets closer or recedes.

For local phenomena, perceived-time $'d\zeta'$ is equal to 'dt' that we will call $'T_o'$; though, being $'T_o{}^{*}{}'$ the absolute-time interval during signal emission:

$$T_o \, / \, T_o{}^{*} = T \, / \, T^{*} = c \, / \, (c \pm v) \qquad (7\text{-}4)$$

Or, in general: $\qquad\qquad \Delta T \, / \, T = v \, / \, c \qquad\qquad (7\text{-}5)$

When the emitter <u>travels away</u>:

$$T \, / \, T^{*} = 1 \, / \, (1 - v \, / \, c) \qquad (7\text{-}6)$$

$$T \, / \, T^{*} = 1 \, / \, (1 - \cos \theta) \qquad (7\text{-}7)$$

When the emitter <u>approaches</u>:

$$T \, / \, T^{*} = 1 \, / \, (1 + v \, / \, c) \qquad (7\text{-}8)$$

$$T \, / \, T^{*} = 1 \, / \, (1 + \cos \theta) \qquad (7\text{-}9)$$

Conventionally, the Doppler's effect relation is written as:

$$\upsilon \, / \, f = (1 - v \, / \, c) \qquad (7\text{-}10)$$

Defining 'z' as: $\qquad z = (\upsilon - fo) \, / \, f = (1 - v \, / \, c)\,) - 1 \quad (7\text{-}11)$

Then: $\qquad\qquad\qquad z = - v \, / \, c \qquad\qquad (7\text{-}12)$

In accordance to the TH principles, it is not advisable to use the term "frequency" when analyzing photons interaction, for several reasons:

The term "frequency" normally refers to continuous phenomena as an ideal sine wave, although absolute time 't' elapses continuously, the length 'T' of photons arriving to the receiver location is not. Referring to frequency may sometimes be useful but is conceptually imprecise. It is frequent mentioning "light" as a continuous wave because of the large quantity of photons involved in most natural phenomena, but when analyzing individual particle interaction, it is very unlikely that the same receiver may absorb continuously similar photons one after another; normally, signal reception is a discontinuous statistical phenomenon since it is almost impossible that two photons follow the same path and additionally to be orderly spaced; most likely, the next photon will have a different period '$T_o$' and will come from a different emitter arriving after a random time interval. Moreover, under natural conditions atomic mechanics will prevent the signal to be absorbed by the same electron in the same atom.

For similar reasons, the conventional term "wavelength" is not appropriate to be used when referring to a particle, a photon cannot be considered a continuous wave and the concept of "spatial distance" is a secondary parameter in a "universe made of time"; nonetheless, to facilitate understanding, we sometimes may improperly use the term "wavelength" or "length of the particle" ($\lambda = c\,T$).

The relative decrease of frequency known as "red shift" in astronomy is what the TH defines as the photon compression:

$$z = - \cos\theta \tag{7-13}$$

$$z = (T - T^*) / T \tag{7-14}$$

Or,
$$z + 1 = T / T^* \tag{7-15}$$

Receiving interaction is always local and as the emitter sub-space is moving with the emitter, the red shift formula does not need relativistic corrections. The TH predicts that no signal can be observed if it is coming from galaxies moving away at speeds greater than 'c'[50], so the usual relativistic correction applied to light waves in astronomy should not be used, because the relative speed of the object under observation is measured locally by decoding the information brought by individual arriving photons; it is irrelevant for the interaction process, that they were emitted by distant objects a long time ago.

---

[50] The observed phenomenon of galaxies that recede at speed larger than 'c', are the consequence of the spatial expansion, as explained in the last chapter of this book.

## Chapter VIII


## Gravity


This chapter analyzes how space cannot exist without gravity; how the universe must be in perpetual motion; how gravity is generated by matter; and how it communicates properties to remote particles. This description ends <u>deducing</u> Newton's "universal gravitational law" and some other related principles.

An ordered universe demands establishing "distance" as the universal reference parameter to base the most fundamental physics laws (gravity, electricity); the remote objects should be less influential than closer ones, but their presence should never vanish; this behavior <u>suggests asymptotic relations</u>.

Moreover, natural order cannot be established unless there is an influence between distant objects in perpetual relative motion; the distance information communicated by asymmetric photons emitted by remote objects that always change relative positions [51] is the key factor to establish order. Without perpetual motion communication, even an initial impulse at the beginning of times would not be able to build space.

Our mental image of a flat space used to define distance and direction for every point with the potential to be occupied by an object, describes an ideal entity not possible to be real. Space as a potential physical entity has a real meaning only when the always static observer receives

---

[51] An always static receiver can perceive distance only when it detects a change in the emitter distance; an always static emitter will not be able to communicate distance.

information that emitters keep changing positions; it will be impossible to define distances in a static universe. Even if all objects are in perpetual motion at constant relative speed, the receiver will detect the motion but cannot decode distance because it will keep receiving photons with the same wavelength; in conclusion symmetric photons can only communicate a fix relative speed, not travelled time, because their length does not change[52], so it is not possible to extract distant information from a parameter that remains constant.

It is important to insist that photon is only able to carry TT information by changing its size; therefore, to define "distance" it is necessary to count on compressed photons that change length as they travel. Remember that photons do not store elapsed TT; they communicate distance indirectly to the receiver when it reads their length; therefore, if all photons would be symmetrical entities they would not be able to communicate distance information. Therefore it can be concluded that gravity is the brick building space.

Current physic theories describe gravity and light as a continuous isotropic radiation emitted by a body residing in a 3D environment; this radiation expands spherically at the speed of light. Consequently its amplitude per unit of area is inversely proportional to the square of sphere radius. But, as analyzed in previous chapters, <u>nature does not communicate by radiating a special substance</u> carrying information; neither decodes distance information by measuring signal's amplitude. In real nature, the particles send information by means of individual photons, and the only way to communicate distance with individual entities is when all emitters move, and the observer could be informed about their individual relative speed. Nature must be in perpetual motion to keep existing...

---

[52] This is not absolutely true for large TT, since photons are entities not having structural auto feedback as the rings; so photons keep changing their length slowly when travelling very long distances generating the Hubble effect. This property fulfills the need for asymmetry to define distances and generate "space" even when gravity becomes negligible or negative.

Before analyzing the gravity, its origin and the way it affects to distant objects, some previous concepts must be revised in full detail.

## The dimensions of energy

The ultimate significance of the term "energy" is related to the proportional effect that the substance characteristics of a particle may affect others by absorption, collision or unbalancing their equilibrium by affecting their symmetry.

Nature stores energy affecting every particular time dimensions: two internal energy components affect the two time dimensions 't' and 'τ' from particle's substance, the Hyle, and there is a third energy component related to the spatial position change of a particle, considering it as an individual entity:

A)   The angular speed defines particle's <u>internal energy</u>; it affects only AT dimension.

B) Any particle loses symmetry when it changes its length n relation with its period; this is a way to store internal spatial energy that could be restored by the action of Hyle elasticity; this energy component affects only internal TT dimension.

C) The material particles as individual entities residing in a spatial location may change their relative position (potential energy) and interchange kinetic energy.

A more detailed analysis on each one of these cases follows:

A)   <u>The internal energy</u> of a particle is directly related to the fact that Hyle is rotating at an angular velocity ($\omega_o = 2\pi / T_o$). Considering that a Hyle structure resembles a rotating helix consecutively pointing to different direction, and that this structure, as a whole, rotates at an angular velocity '$\omega_o$' it has an internal energy depending only on its rotation period 'T*', consequently <u>this energy lies in the dimension of AT</u> since it is defined by only its period 'T*'. Any symmetric material particle is a ring constituted by a directional substance (not independent vectors) rotating on a circular path at an angular speed '$\Omega$' equal to '$\omega$', so their stored energy corresponds to the inertia of the rotating substance (Hyle). Additionally, the whole structure rotates around the center of the ring. Although photons are composed by the same rotating substance, they move permanently in a universal 3D space where all the material particles reside as individual entities circling on their own sub-space; under this perspective the <u>internal energy of symmetric photons</u> is defined only by their Hyle rotation period '$T_o*$', and represents the energy that could be absorbed by rings with high energy.  High energy photons cannot be absorbed; so they collide, and eventually destroy, material particles of equivalent energy.

B) <u>Internal spatial energy</u> relates to the induced change of length of the helix and it is stored as elastic energy. It always <u>affects only the travel-time</u> dimension, the perimeter 'T' of the ring, or the photon's length '$T_o$'. In the case of a ring, any induced asymmetry changes its radius and therefore its inertia, while Hyle angular speed '$\omega$' remains unchanged, or vice versa. This energy component resides in the internal travel-time dimension '$\tau$' of the particle and its changes are ruled by the Hooke's law. In matter particles, the Hyle rotates on itself at an angular speed '$\omega$' while circling the ring at some other angular speed '$\Omega$', caused by the asymmetry producing a metrix change.

C) <u>The external space energy</u> is related to the translation of the ring as a whole, while its Hyle keeps circling on its own-space as an independent static object. Any change in external position is relative to some other referential particle, so it is independent of the internal energy of the particle set by '$\omega$', or the angular speed '$\Omega$' of the ring. This external TT energy can be transmitted to a particle by a compressed object in contact with it or by the kinetic energy transmitted, also elastically, when they collide. It is important to understand that collision means also contact between two independent Hyle structures having similar energy trying to occupy the same point in space; remembering that the most energetic particle will establish the geometry of that point, elastic collision only happens between particles of similar energy, because each one maintaining its individual own-space. The high energy particle will define the straight path in such a way that a weak particle as a photon, will enter to the ring own-space; this is why a low-energy photon does not normally collide with a material particle, it is normally absorbed; on the contrary, high energy photons may destroy a material ring of equivalent energy.

The internal energy contained in a particle remain the same but changes its structure when it mutates by changing residence from a circular own-space to a flat space or vice versa. For instance a ring could be theoretically converted to a photon, by changing its path from circular to linear; this is the case for a photon being emitted by a particle; the opposite will happen when a very high energy photon is emitted by an asymmetric particle, it may wrap up on itself forming a ring. The TH proposes that this is the mechanics of primordial neutrons production, and also in beta emission, this subject will be explained later. The emission process is always originated when a particle becomes asymmetric; it could be a ring with compressed or expanded perimeter.

## Metrix and compression

The absolute time, being the only natural entity which is independent of any other variable, is the only universal reference parameter that can be represented with unconditional accuracy by a linear coordinate axis with constant increment; all other variables are not necessarily linear.

We cannot say the same about TT established by the distance 'T' traveled by a photon during a complete revolution that is defined by 'T'. A compressed photon having (T < T*) increases its TT less than a symmetrical photon having (T = T*).

The TH defines the metrix 'μ' at a point in a sub-space, as the ratio of TT distance 'T' per unit of time (μ = T / T*). When a signal passes a location in space, the spatial metrix is determined by the ratio between the length 'T' and the revolution period of that particular photon. When photons crossing a given point in space have different compressions, the metrix will be computed averaging <u>the statistical directional influence</u> on the receiver during a time interval. A flat space is an abstract entity with metrix equal to one (μ = 1) everywhere and in every direction.

The metrix could be interpreted as the parameter that defines the unit of distance in a spatial location compared with the unit of distance in a flat space, which is used as the universal unit of reference; the AT. This is why the apparent length perceived by a distant observer of an object perpendicular to the line of sight is proportional to the metrix 'μ' at that point. When observing the motion of a remote object; the moving object will be seen as travelling more slowly on a tangential path to the line of sight, when the metrix is smaller than one, and vice versa.

A compressed photon has a length 'T' smaller, than a symmetric photon whose length is equal to 'T*'[53]; this difference in relation to the length 'T' <u>defines compression</u> 'ξ' as the ratio of length decrease (ΔT = T* - T) per unit of length [ξ = (T* - T) / T = ΔT / T]; it establishes the magnitude of the asymmetry of the Hyle structure: it may be a material particle or a photon.

---

[53] It is the same parameter conventionally used in Doppler Effect notation, but with opposite sign because TH considers compression as (T* - T) and not (T − T*); any convention may be established, according to future conceptual convenience.

The metrix '$\mu$' relates to compression '$\xi$' as follows:

$$\xi = \Delta T / T = (T^* - T) / T \qquad (8\text{-}1)$$

$$\xi = T^* / T - 1 \qquad (8\text{-}2)$$

$$\xi = (1 / \mu) - 1 \qquad (8\text{-}3)$$

## Curvature

The mathematical definition of curvature '$\kappa$' of a continuous curve is given by the reciprocal to the radius of curvature 'R', so ($\kappa = 1 / R$). The meaning of curvature in physics refers to a particle moving on a curved path at constant speed, so the angular change 'd$\theta$' of the tangent (and the radius), could be referred to time ($\kappa = d\theta / dt$); therefore; the curvature is proportional to angular speed, '$\Omega$' in the case of a circle.

Applying the same concept to the variation of travel time unit referenced to AT, the space curvature represents the change of direction of distance tensor in a 3D TT environment, so could be defined as a tensor relation between incremental TT and AT, or ($d\tau/dt$). But, remembering that the internal symmetry of a Hyle structure defines the spatial curvature at that point, the curvature in the direction of photon displacement is:

$$\kappa = d\tau/dt = (T^* - T) / T^* \qquad (8\text{-}4)$$

Or in terms of metrix:

$$\kappa = 1 - \mu \qquad (8\text{-}5)$$

Above concept could be applied to external space curvature, and also to the curvature of particle's ring.

## Mass and curvature

It is reasonable to assume that a material particle, whose Hyle establishes that the circle is the straight line in their own sub-space, when releasing tangentially part of its energy, generates a compressed structure with 'T', smaller than 'T*'. Intuitively, the velocity redirection involves a decrease in metrix that is inversely proportional to the perimeter 'T' of the particle, because the rate that direction changes, will be larger for smaller rings. This phenomenon is known as the inertia from a rotating mass in classical mechanics[54].

As it was mentioned before, the curvature 'κ' is the rate of change of the direction ($\kappa = d\theta / dt$) of the velocity vector of a moving object travelling at constant speed on a curved path. Considering that Hyle increases its TT 'τ' at a constant rate ($d\tau/dt$ = constant) while travelling on the particle's circular ring of radius 'R' and curvature ($\kappa = k / R$); then the curvature is inversely proportional to the perimeter 'T':

$$\kappa = k \, d\theta/dt = k \, \omega = k / T \qquad (8\text{-}6)$$

Most material particles in nature, even the asymmetric ones, have 'T' very close to 'T*'; for instance, a free electron has tenths of thousands more energy than a visible light photon; so the curvature of a material particle can be written as:

$$\kappa = k / T^* \qquad (8\text{-}7)$$

The constant 'k' is used along the text of this book as representing a proportionality factor.

The Einstein-Planck relation establishes that mass is also inversely proportional to 'T':

$$m = h / (c^2 \, T^*) \qquad (8\text{-}8)$$

---

[54] The relation between the inertia of the emitters and the gravity generated by them is clearly understood under the TH perspective.

or: $\qquad m = (k\,h\,/\,c^2)\,\kappa$ $\qquad$ (8-9)

Then: $\qquad \kappa = k\,m$ $\qquad$ (8-10)

Therefore, <u>the curvature of a particle is proportional to its mass</u>. This important relation explains how the mass (energy) is the natural consequence of directional change of a rotating substance confined inside its own-space residing as an independent entity. When the Hyle structure (or part of it) changes the geometry of its trajectory from circular to tangential it becomes a compressed photon whose metrix is less than one ($\mu < 1$) as a direct consequence of the geometrical differences between spatial curvatures. This is the way emitted photons have the ability to communicate the magnitude of emitter's mass to remote objects; <u>gravity is originated by this simple fact</u>.

Every part of a photon being emitted increases continuously its tangential distance to the ring, <u>draining part of the emitter elastic substance</u> while the emitter ring keeps rotating; showing that the emission process is a physical phenomenon that cannot be analyzed by applying simple geometric procedures or mathematical conversion theorems.

## Elasticity of rings

The Hyle is the elastic substance constituting the material particles and photons. The origin of its elasticity is the "Law of Symmetry", forcing forces Hyle structures to recover balance by increasing or decreasing its length until reaching symmetry ($T = T^*$).

The Hooke's law defines that the generated force is proportional to the change in particle's length ($F = k\,\Delta x\,/\,x_o = 1 - x\,/\,x_o$), indicating that the force becomes smaller as its length 'x' approaches to its nominal value '$x_o$'. The relative compression refers to ($\Delta x\,/\,x_o$) and the absolute compression to '$\Delta x$'.

Additionally, the substance changes its length slowly because the internal force gets smaller while its geometry gets closer to reach symmetry by transporting mass. The time it takes to reach absolute equilibrium is theoretically infinite since the remaining force decreases as time passes, making the process slower, then ($\Delta t = k / \Delta x$); this equation describes a typical asymptotic process, mathematically defined as a hyperbola (x t = k). In a following paragraph it will be shown how this relation, rules gravitational fields.

Above elasticity analysis applies to any Hyle structure: ring particles or photons.

## **Photon initial compression**

The emission process starts when a ring becomes an asymmetric structure by absorbing energy which causes internal tension that ends up leaking Hyle; this means that the particle recovers symmetry by releasing a small fraction of its energy as a photon.

As mentioned before this process is a physical phenomenon involving a material substance; it cannot be considered simply as a geometric length transformation, only involving the mathematical conversion of distances from a curved space to a flat space or vice versa. The excess of energy released during emission is a continuous process lasting a TT interval equal to the photon's length, '$T_o$'. This process can be visualized comparing it with an elastic filament wrapped up on the rotating ring which is tensioned by the centrifugal force; subsequently, when it leaves the ring tangentially, it decreases its length because it is not subject to the centrifugal force anymore; so the phenomenon could be described in physics terms by considering the photon as a thin rotating ring before the emission which flies away tangentially, converting its rotational energy into elastic energy.

Considering that the subject of the whole process is the substance composing matter (Hyle), which is ruled by the law of conservation of energy; the energy '$\varepsilon_1$' before emission must be the same than the energy '$\varepsilon_2$' after emission.

The total energy '$\varepsilon_2$' of the emitted photon having a period '$T_o*$' and a length '$T_o$', has two components: The internal energy ($\varepsilon_i = h / T_o*$) on the AT dimension and the elastic energy ($\varepsilon_s = k \, \Delta T^2$) stored in the TT dimension where 'k' represents a proportionality factor[55]. This factor is equivalent to a force '$F_o$' needed to compress the total length '$T*$' of the particle ($k = F_o / T_o*$); then, by replacing 'k' we have:

$$\varepsilon_2 = \varepsilon_i + F_o \, \Delta T^2 / T_o* \qquad (8\text{-}11)$$

Or:
$$\varepsilon_2 = \varepsilon_i + (F_o \, T_o^2 / T_o*) \, (\Delta T / T_o)^2 \qquad (8\text{-}12)$$

Considering that ($F_o = k \, T_o*$), then:

$$\varepsilon_2 = \varepsilon_i + k \, T_o^2 \, (\Delta T / T_o)^2 \qquad (8\text{-}13)$$

The relative compression '$\xi$' is equal to ($\Delta T / T_o$); then, assuming that ($T_o \simeq T_o*$); then, the photon energy is:

$$\varepsilon_2 = \varepsilon_i + k \, T_o^2 \, \xi^2 \qquad (8\text{-}14)$$

The initial energy '$\varepsilon_1$', while the Hyle rotates on the ring is defined by Einstein-Planck relation ($\varepsilon_i = h / T_o*$); in this equation the period '$T_o*$' refers only to the amount of energy to be released (photon internal energy) for the ring to regain symmetry. The rotational energy of the ring depends on the ring angular velocity ($\omega_o = 2\,\pi / T_1*$) where '$T_1*$' is the rotation period of the ring, so the portion of energy that releases the ring as a photon is proportional to the square of the angular

---

[55] The factor 'k' in this book is used as a general constant factor whose value depends on the system of units used. When appearing on both sides of an equation, it will represent different constant values. The value of 'k' in the elastic force ($F = k \, \Delta T$) is the ratio between the theoretical force '$F_a$' needed to compress completely the particle ($k = F_a / T*$), assuming linearity. When the particle has a small compressions it can be written as ($k = F_a / T$)

velocity ($\varepsilon_r = k\,\omega_o^2$), which is proportional to the total energy from ring:

$$\varepsilon_1 = \varepsilon_i + k\,\omega_o^2 \qquad (8\text{-}15)$$

Then:
$$\varepsilon_i + k\,T^2\,\xi^2 = \varepsilon_i + k\,\omega_o^2 \qquad (8\text{-}16)$$

$$T^2\,\xi^2 = k\,\omega_o^2 = k\,/\,T_1^{*2} \qquad (8\text{-}17)$$

$$\xi\,T = k\,/\,T_1^{*} \qquad (8\text{-}18)$$

Or:
$$\Delta T = k\,/\,T_1^{*} \qquad (8\text{-}19)$$

The initial relative compression '$\xi_i$', when the structure leaves the emitter, is:

$$\xi_i = \Delta T_i\,/\,T_i \qquad (8\text{-}20)$$

Considering that the rotational energy of the just released photon '$\varepsilon_r$' is a fraction of the ring's energy, as both rotate at the same angular speed ($\omega_o = 2\,\pi\,/\,T_1^{*}$), this energy is proportional to the particle's mass ($\varepsilon_r = k\,\varepsilon_o = k\,M$). In other words, since the mass of a particle is defined as ($M = h\,/\,c^2\,T_1^{*}$), so the term ($k\,/\,T_1^{*}$) from equation (8-19) is proportional to mass 'M' of the emitter; therefore, <u>the initial compression '$\xi_i$' is also proportional to the mass 'M' of the emitter</u>; then:

$$\Delta T_i = \xi_i\,T_i = k\,M \qquad (8\text{-}21)$$

It will prove to be convenient to write above equation as:

$$\xi_i\,T_i = K_g\,M \qquad (8\text{-}22)$$

Or, in terms of absolute compression '$\Delta T_i$':

$$\Delta T_i = K_g\,M \qquad (8\text{-}23)$$

The factor 'K$_g$' is a natural constant related to the gravitational universal constant G; having the same magnitude when using the natural system of units.

Consequently, the compression is also proportional to particle's ring curvature 'κ'.

$$\Delta T_i = \xi_i \, T_i = K_g \, M = k \, \kappa \qquad (8\text{-}24)$$

It is important to note that the absolute compression of the emitted photon is proportional to emitter's mass, producing a variable relative compression that is a function of its length; giving as result that the travelled distance per unit of absolute time is independent from emitted photon's length.

### Relative speed detection

A photon communicates status information on the emitter; its existence, its charge, polarity, and mass. Considering that in any multidimensional spatial environment, the magnitude of the influence of mass and electric are functions of distance, the photons must have the capability to somehow communicate distance information or TT, by decreasing its effect when distance increases.

The only way for photons to communicate travelled TT is indirectly, by changing their length as they travel away from the emitter; it is obvious that only asymmetric photons may do it as the result of the internal force generated by the "law of symmetry"; a symmetrical photon cannot provide information on the distance to the transmitter so they are not capable to communicate attraction or repulsion.

The photon <u>emitted by a free particle</u> carries emitter information on its charge and mass; this is because the photon compression is a function of the ring size and its structure is set by the ring architecture. Considering that free particles are normally free charges (as electrons or protons) communicating electrical forces much larger than gravitational forces [56] the mass of the free emitter is irrelevant; mass information is only relevant when signals originate from large atomic structures.

In a previous paragraph has been shown how relative motion is decoded by the receiver, and how the relative speed of the emitter is related to the perceived relation between the length 'T' of arriving photon and its period 'T*', given by the equation:

$$\Delta T / T = v / c \qquad (8\text{-}25)$$

The length of a compressed photon (T < T*) will be interpreted as that the emitter is approaching at the relative speed given by above equation.

**Space as a velocity field**

Velocity is defined as the travelled distance 'd$\tau$' per unit of AT (v = d$\tau$ / dt); it is the parameter that nature directly communicates to the receiver. But previously we must remember that the photon travelled distance 'T' per revolution establishes the unit of length in that location, and that the communicated speed depends on the photon compression. When an object emits asymmetric photons which change compression as time passes, the spatial metrix will change accordingly; therefore, the speed informed to the receiver <u>is really the relative speed</u> at which the emitter moves, because it can also be interpreted as <u>the speed at which the subspace moves in that location</u>.

It is understandable that this concept is difficult to digest to our mind accustomed to use "flat-space" as reference, but we must remember that <u>"distance" is an abstract parameter</u> set by the photon as absolute

---

[56] Free electron or proton, are not relevant sources of gravitational attraction. A free neutron is a sui generis inactive particle in the material universe; the gravitational influence of its mass is also irrelevant in a macro world with abundant complex molecular structures.

time passes, and that physical reality is that distance is a mental image of another intangible parameter: the travelled time.

The change of the spatial metrix as described above, is what the TR denominates "space curvature" referring to it as the variation of metric unit at different locations; now we can understand that is not absolutely appropriate to say that "massive bodies bend space", instead of properly stating that "space curvature is an abstract concept of distance unit established by elongating photons"; so there is a profound difference: for TR space is a substantial physical entity and for TH a contingent mental abstraction of a natural phenomenon.

A flat space is just an unreal entelechy and cannot exist in nature because symmetric photons cannot communicate distance information; consequently, it is reasonable to affirm that curved space is the only natural space[57]; we might say that gravitational sub-spaces are the only "real spaces", always generated by a gravitational field instituted by a material object emitting compressed photons.

This model describes a gravitational sub-space as an environment where every location is related to a speed; under this perspective, space should be properly defined as a contingent entity that could be mathematically <u>described as a "velocity field"</u>; so, a <u>natural "sub-space" is not a constant metric entity</u>. Consequently, <u>gravity generates a natural sub-space</u> when massive bodies establish the spatial properties of vicinity by continuously emitting photons.

The best way to didactically describe a 2D sub-space G is to imagine a large flat container with water, with a drain on its center. The water surface will travel at higher speed when closer to the drainage, forming circles of constant velocity around it; the size of the outlet will

---

[57] It will be shown later that intergalactic space is an extension of this argument; a negative curvature space is necessary for communicating the "distance" when gravity vanishes.

establish the velocities, so it will represent the mass of the emitter. If there is a cork floating on the water surface, it will accelerate in the drain direction, since it will increase its speed when approaching to the hole. Any material object, a lead ball or a feather, will be as the cork floating toward the drainage because it is the surface the one that moves, dragging the object.

The same 3D model fitting more closely the real phenomena may be described as a vacuum cleaner that drains the smoke filling a large room; every spherical surface around it will represent a constant velocity surface, and the power of the vacuum cleaner will define the mass of the emitter; so, any floating feather will accelerate toward the absorption point.

**<u>Composite space metrix</u>**

The photons coming from the same particle will never collide or affect each other. Photons emitted by a larger source may do it but most of them are ahead or behind each other, and they increase their TT at the same rate because they have the same compression.

A photon is a one dimensional entity following a straight path that may cross simultaneously the path of another colliding with it; considering that both particles have similar energy content, they cannot absorb each other. The collision probability depends on the relative quantity of photons emitted by each source; for instance, the few photons arriving from a distant star will collide, and modify their paths, when crossing the vicinity of a massive bright emitter, as the sun, that produce myriads of photons travelling in every direction. Any photon intercepting the radial path will tend to follow the radial current similarly to fluid molecules flowing together.

This ensemble builds spherical "spaces" around each emitter, so its addition generates a composite statistical entity causally related to neighboring masses; the metric properties in any point in "space" are set not only by each individual photon metrix but also by the relative quantity of photons arriving to that point from every source in the vicinity; the "properties" in that location will be stable as long as the same relative quantity of photons keep arriving in the same proportion from each source.

## **Photon decompression**

As previously shown, any photon length 'T' becomes closer to 'T*' as its TT '$\tau$' increases '$\Delta\tau$'. Unit distance will be defined by length 'T', so the TT is a discrete parameter where ($\Delta\tau = T$). Additionally, the relative compression of the photon ($\xi = \Delta T / T$) will define the metrix at every point of the subspace it crosses, and will apply the same relative internal force to photon's helix of any length, as ruled by Hooke's law; consequently all photons expand at the same relative rate.

The compression of a Hyle helix is defined as a relative value referred to the length; valid for any photon size. Then, the equation describing the phenomenon is:

$$\xi\,\tau = A_g \tag{8-26}$$

Where $A_g$ is a constant whose value can be found when knowing the photon initial compression '$\xi_i$' when leaving ($\tau = T_i$) the emitter with mass 'M'; in this way, the factor '$A_g$' is proportional ($K_g$) to emitter's mass 'M'. This is because the magnitude of '$A_g$' is given in TT units, so one must refer it to the period '$T_o$' of the photon as the TT unit of reference[58], in the quotient ($A_g / T_i$), the period '$T_i$' will always be one ($T_i = 1$):

---

[58] From a physical perspective, this equation establishes the identity between to concepts: a) the photon asymmetry originated by the compression ($\Delta T$) in relation to its length 'T', and b) the "average spatial asymmetry" given by a relative compression factor '$A_g$' in relation to the travelled time '$\tau$'; both being numerically identical at the moment of emission.

$$A_g = \xi_i\, T_i = K_g\, M \qquad\qquad (8\text{-}27)$$

Then the general equation ruling photons decompression phenomena is:

$$\xi\, \tau = K_g\, M \qquad\qquad (8\text{-}28)$$

That can be written as:

$$\Delta T / T = K_g\, M / \tau \qquad\qquad (8\text{-}29)$$

This relation indicates that initial compression will be defined by the emitter mass 'M' since '$K_g$' is a universal constant as experimentally verified. Usually the emitter is an object composed by a large quantity of diverse molecules emitting photons with different energy but sharing the same relative compression and therefore metrix '$\mu$'; consequently, the spatial distribution of compressions will depend on the magnitude of the emitter mass, and the remoteness of the receiver.

Since ($\Delta T_i = K_g\, M$), this equation could be written as:

$$\Delta T\, \tau / T = \Delta T_i \qquad\qquad (8\text{-}30)$$

The ratio ($\tau / T$) is the number 'n' of Hyle complete revolutions (n = $\tau$ / T); so the above equation could be expressed as:

$$\Delta T = \Delta T_i / n \qquad\qquad (8\text{-}31)$$

This relation tells us how the photon compression vanishes as it travels away from an emitter which established its original compression '$\Delta T_i$' proportionally to its mass 'M'. This simple equation reveals the most basic law that rules the creation of space and gravity, and its discrete condition. It is overwhelming how a simple numerical series of natural numbers describes the fundamental natural law constituting one of the cornerstones supporting the universe.

## The universal gravitation law

A receiver located within a "Gravitational Sub-Space" (sub-space G) absorbs compressed photons that increase their 'T' as they travel away from the emitter in accordance to the relation[59]:

$$A_g \, / \, \tau = \Delta T \, / \, T \qquad\qquad (8\text{-}32)$$

This equation can also be written as:

$$A_g = (\Delta T \, / \, T) \, \tau \qquad\qquad (8\text{-}33)$$

The factor '$A_g$' is measured in time units, and it defines the Hyperbola focal position and it is not a constant as it depends on emitter's mass. The compression informs to the receiver that the emitter is approaching at a relative speed, because $(\Delta T \, / \, T)$ represents the relative velocity 'v' given by Doppler's effect relation $(v \, / \, c = \Delta T \, / \, T)$, and $(A_g = K_g \, M)$; then:

$$v \, / \, c = K_g \, M \, / \, \tau \qquad\qquad (8\text{-}34)$$

Therefore, the <u>subspace G represents a velocity field</u>, so the velocity change 'dv' in relation to a position change in 'd$\tau$' space will define an acceleration field 'g', which is conventionally referred as "the acceleration of gravity"[60]; then:

$$g = dv/d\tau \qquad\qquad (8\text{-}35)$$

---

[59] It is important to note that the TT reference used in this equation is the photon's period 'T', which is not a constant since it changes value with position, so it is not an absolute reference; so both ratios in this equation are relative terms.

[60] The classical definition of acceleration is $(dv/dt)$, therefore deriving the velocity with respect to position is only approximate since presupposes that $(dt \sim dt)$ which is only valid for small gravity values $(\xi = \text{constant})$. The right relation should be $(a = dv/dt = \xi \, dv/d\tau + v \, d\xi/d\tau)$.

Then: $$g = - c \, K_g \, M / \tau^2 \qquad (8\text{-}36)$$

Converting TT '$\tau$' to metric distance 'r' by applying the conversion factor 'c', we find (r = c $\tau$), so:

$$g = - c^3 \, K_g \, M / r^2 \qquad (8\text{-}37)$$

This relation turns out to be the well known Newton's Universal Gravitation Law:

$$g = - G \, M / r^2 \qquad (8\text{-}38)$$

Then the constant '$K_g$' relates to the universal gravitation constant 'G':

$$K_g = G / c^3 \qquad (8\text{-}39)$$

In the "natural system of units" where (c = 1); then (Kg = G).

It is very important perceive that we have deduced the universal gravitation law and the existence of 'G' as a constant, by following only logic procedures based on the TH proposals.

The gravitation constant 'G' then is:

$$G = K_g \, c^3 \qquad (8\text{-}40)$$

And the velocity field in a gravitational sub-space (gravity potential), in a conventional form is given by:

$$v_g = G \, M / c \, r \qquad (8\text{-}41)$$

**<u>Gravitational red shift</u>**

The observed "red shift of light" as the distance from a massive body increases is conventionally interpreted as the effect of gravity on light frequency; explaining that the "gravitational pull" that also travels at light speed also has the same effect on light (and photons), but not explaining how is possible that gravity travels together with light and influence it. The TH demonstrates that the opposite is the truth; photon compression is the cause, and gravity the effect.

$$\Delta T / T = K_g \, M / \tau \qquad (8\text{-}42)$$

Then:
$$T = T^* / (1 + K_g \, M / \tau) \qquad (8\text{-}43)$$

Converting TT to metric distance ($r = c\,\tau$):

$$T = T^* / (1 + G\,M / c^2 \, r) \qquad (8\text{-}44)$$

In conventional notation, where ($\lambda = c\,T$):

$$\lambda = \lambda_o / (1 + G\,M / c^2 \, r) \qquad (8\text{-}45)$$

Therefore, the gravitation phenomenon is the result of photons red shift; or, the phenomenon known as "gravitational red shift" <u>confirms that light generates gravity</u>.

Under the TH perspective, it is obvious that any matter concentration that cannot emit photons, like the proposed mathematical entity known as black hole, will not build a gravitational subspace; additionally it has to be an extremely cold body with no internal signal interchange not receiving any external signal (being very far from any other emitter); in short, it will be an entity that does not communicate and consequently it will not be part of our universe.

The last equation could be written in reference to metrix '$\mu$' as:

$$\mu = 1/(1 + G\,M / c^2 \, r) \qquad (8\text{-}46)$$

Or in TT terms:
$$\mu = 1/(1 + K_g \, M / \tau) \qquad (8\text{-}47)$$

This equation describes the metrix distribution in a gravitational subspace.

## Force and space

The basic laws of mechanics relate forces with motion, so it is important investigate the origin of force, its meaning, and its relation with space.

The term "force" in physics refers to a physical parameter closely related to matter elasticity that is the natural property of matter oppose to any change induced to its original size; this force is internally generated by asymmetric particles in observance to the "law of symmetry" ruling that the length 'T' of a particle (measured in TT units) must be equivalent to its period 'T*' in AT units. Under this local perspective, the concept of "force" as defined by the TH only applies to the internal elasticity of individual Hyle structures that cannot affect directly to other material entities residing in distant locations.

On the contrary, for classical mechanics "force" is a concept relating spatial positions, time and mass; but, it was demonstrated in previous pages that external space is an abstract entity whose properties are defined by the characteristics of the signals crossing it; consequently it is not possible to directly relate the laws governing motion with equations involving internal particles behavior[61].

It was demonstrated previously that acceleration is independent on the mass of the object residing at a subspace G. The term "space" represents the image of a velocity field that can be perceived also as a gravitational field; this abstraction is coincident with the classical perception.

To analyze the relation between external space and internal elasticity consider an emitter "O" generating a subspace G, a receiver placed at

---

[61] It is important to note that by relating force to spatial position change, one automatically assigns elastic properties to the abstract entity "space."

point "A", and a mass 'm' at point "B" closer to "O". Let's suppose that each one of the particles at "B" is tied to the receiver at "A" with an elastic filament with a length '$\tau_a$'. The object at "B" will receive photons with smaller metrix than the receiver at "A", so "B" approaches the emitter at higher velocity than the receiver. Since the receiver is the always static natural reference point, the velocity increment '$\Delta v$' represents that the acceleration 'a' is pulling "B" toward the emitter[62]; we will consider this acceleration be constant.

Every filament will increase its length by '$\Delta T$', generating a traction force given by Hooke's law as ($F = k \, \Delta T \, / \, \tau_a$) attracting every particle to "A"; both effects, the space pulling apart and the attraction of the filament, will balance in a position close to "B". Consequently, the force 'F' will be proportional to the acceleration of every particle that receives a photon at "B" ($F = k \, a$), and the total force 'F' applied to the object at "B" will be proportional to the number 'n' of individual particles constituting it[63]; as the mass 'm' is proportional to 'n', the force will be proportional to the mass 'm' and to the acceleration 'a'; so:

$$F = k \, m \, a \qquad\qquad (8\text{-}48)$$

The value of 'k' is given by the system of units used.

---

[62] When "B" does not move relative to "A" there is no need to apply a force to stop "B", and the same will happen if "B" has a constant relative velocity and it travels on a flat space; consequently the force must be proportional to the increment of relative velocity. When "B" moves in a subspace G, it changes relative speed because it resides in a velocity field, so moving means accelerating. Additionally, an object changing "position" in a velocity field relates changes in "distance" with change of "velocity"; and a change in distance is related to "force" in an elastic media, this is why "force" relates to "acceleration", relating the concepts of "elasticity" and "space."

[63] Not all the particles from a complex object containing billions of them will receive the emitter signals, only the ones at the surface facing it; but as it will be shown, the metrics changes are communicated to all the particles in a body, because their atoms and molecules are elastic structures in permanent communication.

The same result will be found if the filaments and the mass at "B" are replaced by any other device applying an equivalent force preventing it moving with respect to "A"; it could be a solid rod or a motor. In conclusion, any object residing in subspace G will accelerate toward the emitter independently from its mass, unless an opposite force compensates the gravity pull; this force is proportional to the mass and to <u>the relative acceleration perceived by the receiver</u>.

The acceleration is always relative to the observer, and it may be originated by a process other than gravity. The force needed to stop a relative acceleration of object in relation to the receiver will induce an equivalent acceleration to rest, relative to the receiver. Therefore, the term "force" always represents a natural action equivalent to the stress that any material object generates when it changes size; is always originated by the internal energy in the TT dimension, and may induce relative motion; or a position change in the external TT dimension[64].

Individual matter particles always maintain internal symmetry unchanged; <u>internal unbalance</u> producing "forces" is always generated by external influences as:

- Photon absorption disturbs the internal balance of the particle; this is the case of internal force generated in a receiver located on a gravity field. Unstable particles are short lived because they have an initial unbalanced structure generating large internal elastic forces breaking them.
- Complex emission occur when atoms and molecules reconfigure their internal geometry by a chemical or a nuclear process.

---

[64] From the TH perspective, the equivalence of inertial mass and gravitational mass and that gravity is a relative acceleration, is so obvious that does not need to be clarified.

- Elastic forces generated by collision.

## Particle clusters

The interaction among particles demands stable communication; this prerequisite can only be fulfilled among large clusters of particles. A solitary particle or a small group of them cannot exchange signals permanently with the outer environment; consequently, they do not have a permanent presence in the universe. Just imagine a lonely electron emitting a photon directly to a receiver which absorbs it and sends it back in exactly the same direction; of course this is highly improbable in nature and certainly not recurrently. The natural condition imposed by the fact that signals travel in a one dimension path, within a 3D environment, automatically determines that the communication process, and the abstract entities defined by it, be statistical phenomena.

An effective emitter must involve large number of particles sending signals permanently in all directions. Massive objects at high temperature, permanently communicate elastic energy between their atoms as it will be explained later.

A single receiver cannot absorb signals indefinitely; its storage capability is limited, so the receiver must not only contain many particles, but its structure must be capable of absorbing a wide range of energy; the energy from particles on the surface of a material body is elastically transferred, by so increasing molecules mobility (temperature), the internal metrix (potential energy) and motion. As it will be shown, nature solves these requirements by structuring atoms and molecules with wide range of energy absorption capabilities.

## Chapter IX

## Electricity

It is universally accepted that charges are individual particles that interact remotely by applying a force which is inversely proportional to the square of the distance separating them.

This chapter shows that the internal structure of a particle defines it as a positive or negative charge, and that the individual emitted photon always defines both; the metrix of the sub-space G and the magnitude of the electric force applied to the receiver.

It also shows why the electric field is a force field, while the gravitational field (or "space") is a velocity field.

### Electric Sub-space

The TH assumes that the electric properties of any material particle are given by its internal structure. One of the three possible structures described in previous chapters defines neutral particles, a second one corresponds to positive charges, and the third one necessarily identifies negative particles; there is a fourth transitional state corresponding to asymmetric particles that lost their static status; this unstable configuration may manifest momentarily positive or negative electric qualities until they liberate energy excess in the form of a photon when reaching a stress limit. The so defined "charged particles" emit a polarized photon (Pp) [65] with internal structure $A^+$ or $A^-$; the neutral particles will emit normal photons (Np) of the type $B^+$ or $B^-$.

---

[65] This assertion will be proven in chapter XII

The gravitational space is generated by normal photons (Np) absorbed by particles, with or without charge, while an electric sub-space is generated only by charges. Then, the electric phenomenon may be explained by one of the following alternatives:

a) Assuming that polarized photons "Pp" initial compression is defined by the emitter structure independent of emitter mass, so generating an independent subspace E having larger velocities (lower metrix) on the spatial location set by a sub-space G. In this case, an external observer residing in the sub-space G will perceive that a Pp travel slower than a Np, communicating larger asymmetries to receivers[66]. This assumption is erroneous because the emitter mass defines the emitted photon compression; obviously emitters do not have a different mass for each direction of rotation of its Hyle disc (plane of rotation); additionally, the emission mechanics shows that the inertia of the emitter is only a function of the angular velocity of the ring '$\Omega$'; consequently, is independent of the orientation of Hyle vectors plane of rotation.

b) That Np and Pp photons have the same initial compression defined by the mass of the emitter, and it is only the receiver internal structure which generates an appropriate force, depending on the polarity of Pp absorbed. Under this alternative is reasonable assume that the Hyle of a Pp has its angular velocity vector '$\omega$' pointing towards or against photon's displacement direction, denote sign, while an Np always has '$\omega$' perpendicular to photon axis; this explanation will be analyzed in full deepness in chapter XII.

---

[66] This erroneous assumption has been assumed in our first book, where the sub-space E was described as having other metrix than sub-space G.

Considering that the second alternative is more coherent with observed behavior of photons, it is reasonable assume that the decompression affecting a Pp moving away from the emitter is driven by internal elasticity force ruled by the "law of symmetry" described by the same asymptotic criteria as in the case of gravity.

Then, both the Pp and the Np have the same absolute initial compression ($\Delta T = \xi_i \, T_i$) defined by the mass 'M' of the emitter ($\xi_i \, T_i = K_g \, M$); therefore, the metrix of both: sub-space G and sub-space E, is the same ($\tau = \tau_e$); consequently, the velocity field '$v_e$' and the acceleration field '$a_e$' may be described by a similar equation:

$$v_e = A_e \, c \, / \, \tau \qquad\qquad (9\text{-}1)$$

The assumed electric velocity field used in this equation does not describe an independent sub-space, but is the name given to it in the present analysis considering that the electric force applied to a charge accelerates it and changes its velocity on each point; this velocity always differ substantially from the velocity set by a sub-space G, because the acceleration '$a_e$' is generated by a force '$F_e$' applied to the receiver mass 'm', defined as [$a_e = (dv_e/d\tau)$]; then:

$$a_e = A_e \, c \, / \, \tau^2 \qquad\qquad (9\text{-}2)$$

The parameter '$A_e$' will derive from the magnitude of the applied electric force that overcomes the object tendency in adopting the velocity set by the sub-space G; or in conventional language, *it must defeat the natural resistance of its mass to travel on its inertial path.* Consequently, the generated acceleration '$a_e$' in sub-space G equal to the force per unit mass ($F_e \, / \, m = a_e$). The magnitude of this force was found experimentally and is given by the Coulomb's law ($F_e = k_e \, q^2 \, / \, r^2$), where 'r' is the distance in accordance to the distances and metrix set by the sub-space G; therefore:

$$F / q_1 = - k_e \, q_2 / r^2 \qquad\qquad (9\text{-}3)$$

The magnitude of a unit charge 'q' is a binary variable, because it defines if the receiver has a specific structure or not; so the magnitude of 'q' will depend on the system of units used[67]; but conceptually must be dimensionless and equal to one with a sign. Therefore both, the emitter and the receiver will always have the same unitary charge ($q = q_1 = q_2$); then, for a receiver with mass 'm':

$$m \, a_e = k_e \, (q / r)^2 \qquad\qquad (9\text{-}4)$$

Or:
$$a_e = (k_e / m) \, (q / c \, \tau)^2 \qquad\qquad (9\text{-}5)$$

This acceleration created by the electric force generates a "force field", described as a powerful sub-space E which applies strong forces only to charged particles (or charges). While the sub-space G is a velocity field (or acceleration field) where the velocity (or acceleration) is caused by a "moving space"; the electric field is NOT a velocity field, but a force field, which applies a force on a charge on it; this force is proportional (much greater) to the force assigned to the sub-space G ($F_g = m \, a_g$), so induces a much larger acceleration ($a_e = F_e / m$) to the particle by using the internal elasticity of the receiver to move it[68]. Consequently, the sub-space E will be a "force field" which corresponds to an "electric acceleration field" described as ($a_e = A_e \, c / \tau^2$).

Replacing this '$a_e$' in equation (9-5), we have:

$$A_e \, c / \tau^2 = (k_e / m) \, (q / c \, \tau)^2 \qquad\qquad (9\text{-}6)$$

$$A_e = (k_e / c \, m) \, (q / c)^2 \qquad\qquad (9\text{-}7)$$

---

[67] In natural system of units ke = 1/[sec]; and q = [].

[68] The details of this phenomenon will be analyzed later.

This equation describes the specific case of an individual emitter with a charge 'q' acting on a particle with mass 'm' and charge 'q'. This equation explains that '$A_e$' is inversely proportional to the total mass 'm' of the receiver ($A_e = k / m$); the receiver may be an individual particle with mass 'm' or a complex structure with mass 'm' where the charge resides. Considering that the total effect is additive, a composite sub-space E for each specific emitter-receiver interaction can be described by an statistic entity involving the vector addition of all individual charges residing on an emitter and the group of charges in the receiver.

The above equation includes only one variable, the mass 'm' of the receiver, while all other factors are constants; then:

$$A_e = k / m \qquad\qquad (9\text{-}8)$$

## Sub-space E metrix

It has been shown that the metrix of the sub-space E is exactly the same than the metrix of sub-space E.

The "electric force field" or sub-space E will produce an acceleration [$a_e = (k_e / m) (q / c\,\tau)^2$], which may be assigned to an abstract velocity field given by:

$$v_e = (k_e / m) (q^2 / c\,\tau) \qquad\qquad (9\text{-}9)$$

Whenever a sub-space E is mentioned in this text, it will refer to the velocity field given by this equation.

## The fine structure constant '$\alpha$'

Multiplying both sides of the equation (9-7) by ($2\pi / \lambda_\varepsilon$), where ($\lambda_\varepsilon = c\,T_\varepsilon$) is the wave length of the electron or the perimeter of its ring; then:

$$(2\,\pi\,/\,\lambda_\varepsilon)\,A_e = 2\,\pi\,(k_e\,q^2)\,/\,(m\,c^2\,\lambda_\varepsilon\,c) \qquad (9\text{-}10)$$

In accordance with the Einstein-Planck equation, the internal energy of the electron is:

$$m_\varepsilon\,c^2\,\lambda_\varepsilon = h\,c \qquad (9\text{-}11)$$

$$(2\,\pi\,/\,\lambda_\varepsilon)\,A_e = 2\,\pi\,(k_e\,q^2)\,/\,(h\,c^2) \qquad (9\text{-}12)$$

The "fine structure constant" '$\alpha$' is defined as:

$$\alpha = 2\,\pi\,(k_e\,q^2)\,/\,(h\,c) \qquad (9\text{-}13)$$

Then: $\qquad\qquad A_e = \alpha\,\lambda_\varepsilon\,/\,2\,\pi\,c \qquad (9\text{-}14)$

The radius '$R_e$' of electron ring is:

$$R_{\varepsilon} = \lambda_e\,/\,2\,\pi = c\,T_e\,/\,2\,\pi$$

$$A_e = \alpha\,T_e\,/\,2\,\pi \;\; = \alpha\,R_e \qquad (9\text{-}15)$$

Similarly, when the charge is a proton:

$$A_p = \alpha\,T_p\,/\,2\,\pi \;\; = \alpha\,R_p \qquad (9\text{-}16)$$

By applying the Doppler relation ($v\,/\,c = A\,/\,\tau = \Delta T\,/\,T$) to a photon with period 'T' generating a sub-space G, it was found that ($A_g = K_g\,M$). The above equations show that '$\alpha$' is the constant in the sub-space E, similar to the universal constant '$K_g$', and the radius '$R_e$' or '$R_e$' ($R = T\,/\,2\,\pi$) resembles the emitter mass 'M'. $A_g$ was the initial absolute compression, defining the initial velocity value ($v = c\,\Delta T\,/\,T$) for the gravity "velocity field"; in this case, '$\alpha$' is a constant, and the radius 'R' of the particle's ring assign an initial velocity for the abstract concept of sub-space E.

## Gravity force and electric force ratio

It has been shown before, how a charge emits a Pp that communicates an attractive or repulsive force to a charged receiver depending on its polarity, and that the Pp applies simultaneously the gravitational force given by the magnitude of its compression and the electric force defined by its structural architecture.

Is very important remind that the magnitude of the electric force also depends on the distance, and that distance information is provided by the compression of the photon that also sets the metrix of the sub-space G; in conventional language we may say that an electric field <u>always resides on a gravitational field</u> because it is generated by the same particle emitting a photon with similar compression, so an electric field <u>cannot exist as an entity independent from a gravitational field</u>. Individual charges communicate always gravitational information by emitting compressed photons, whose compression is defined by its mass, while its structure is defined by its polarity.

Therefore we must always keep in mind that is the same photon the one which carries the electric and the gravity information; therefore, the metrix of the sub-space E is the same than the metrix of the sub-space G because the receiver decodes the asymmetry of the photon as the distance information for both phenomena.

The electric phenomenon can be visualized as the interaction between a screw (photon) rotating as it travels, and intercepts a static gear (the receiver); depending on the linear velocity, and the relative magnitude and direction of the angular velocity '$\omega$', the gear will rotate toward or away from the emitter. On the contrary, when the same screw reaches a toothless wheel, it will not induce rotation on it.

It is important understand that the electric force is always exerted by a single photon on an individual particle; the overall electric effect on a body that contains charged particles is found by integrating the effect on all of them. It is also important to mention that a charge (free or inside a body) emits one Pp after absorbing a photon, therefore no charge emits Pp continuously; this is why, there are few Pp emitted by just a few charges in a massive body as compared to the several orders of magnitude higher of Np emitted by the same object; additionally not always the same charge will send a Pp to the same receiver, consequently, the average electrical effect is negligible compared to the consolidated gravitational action from millions of neutral particles. This is the reason that artificial geometrical arrangements are needed to take advantage of the forces generated by electric phenomena.

The relationship between electrical '$F_e$' and gravitational force '$F_g$' exerted between free particles is:

$$F_e = k_e \, q^2 / (c \, \tau)^2 \qquad (9\text{-}17)$$

$$F_g = G \, M \, m / (c \, \tau)^2 \qquad (9\text{-}18)$$

$$F_e / F_g = k_e \, q^2 / G \, M \, m \qquad (9\text{-}19)$$

$$\alpha = k_e \, q^2 / h \, c \qquad (9\text{-}20)$$

Then: $\qquad F_e / F_g = h \, c \, \alpha / G \, M \, m \qquad (9\text{-}21)$

Substituting ($G = K_g \, c^3$):

$$F_e / F_g = h \, \alpha / K_g \, M \, m \, c^2 \qquad (9\text{-}23)$$

The photon absolute compression '$\Delta T_i$' is defined by the emitter's mass '$M$' as:

$$K_g \, M = \xi_i \, T_i = \Delta T_i \qquad (9\text{-}24)$$

Then: $\qquad F_e / F_g = h \, \alpha / \Delta T_i \, m \, c^2 \qquad (9\text{-}25)$

The receiver with mass 'm' has a period '$T_r$' given by the relation ($T_r = h / m c^2$); then:

$$F_e / F_g = \alpha \, (T_r / \Delta T_i) \qquad (9\text{-}26)$$

This equation indicates that the ratio of the electrical force and the gravitational force exerted on a receiving particle (usually an electron) is proportional to the receiver's period '$T_r$', suggesting that the electric force is generated by modifying the internal structure of the receiver when it absorbs an asymmetric photon, causing the ring deform and move its center while recovering its symmetry. The equation also indicates that the gravitational force, in relation to the electric force, increases when the <u>emitter mass increases</u>, therefore decreasing the relative effect of the electrical force. This compression is minimal for free emitters (electron or proton); so, the gravity force is irrelevant for free charges.

**This page is intentionally left in blank**

## Chapter X

## Magnetism

Some classical theories consider magnetism as a primary natural parameter; this perception is based on the experimental observations on the forces generated by distant magnets or electric currents. In the analysis that follows it will be shown that magnetism is just a force generated when an object moves on non inertial trajectories. Consequently, magnetism in NOT a substantial natural parameter, but just the effect on moving charges. One may conveniently use an "intermediate variable" 'B' simplifies the analysis of the complex geometrical interaction among moving charges.

Before analyzing the magnetic phenomena in deep, some previous concepts should be clarified.

### Circular motion in a sub-space

The term sub-space refers to an abstract entity continuously generated by its emitter, even when it moves, the sub-space will "move" with it, because photon's motion is always referred to an emitter that remains static. A receiver moving freely at slow speed in a sub-space G will end up describing a path shaped as by a conical curve, because the centrifugal radial component of its velocity compensates the centripetal radial velocity of the sub-space; in the simplest case, it travels at a constant tangential velocity on the circle of constant radial velocity.

A "free receiver" is not subject to external forces other than gravity; it does not have a self-propulsion motor; so, it will always follow what is conventionally known as an inertial path. All initially static material objects, describe an inertial path as they move at the speed and direction imposed by the composite-space on which they reside.

This important property of a multi-dimensional space allows a receiver move freely on a two dimension plane by following a conical trajectory reaching an appropriate tangential velocity whose incremental radial displacement compensates the rate of approach imposed by the radial velocity field. In general tangential motion in central force fields is the mechanism used by nature for keeping material objects separated from the emitter.

**<u>Remotely observed motion</u>**

An observer perceives distorted motions from objects moving on perpendicular paths to the line of sight inside a distant sub-space. He perceives slower motions and apparent shorter distances when the moving object is in a sub-space with smaller metrix, similarly when observing trough water, since in both cases the observer perceives phenomena through a transparent media inside a sub-space with fractional metrix. This statement is limited to objects moving at slow speeds, so don't need relativistic correction.

For a remote observer, a constant tangential speed ($v = ds/dt$) means constant ($d\theta/dt$) where ($ds = r\,d\theta$) and ($r \gg s$).  When the metrix of its path is less than one ($\mu < 1$), from the observer's perspective the distance 'ds' is shorter ($\mu\,dl = ds$) and the apparent speed is slower because the angle's increment ($d\theta$) decreases:

$$ds/dt = \mu\,dl/dt \qquad\qquad (10\text{-}1)$$

This is observed when seeing distant celestial objects. This phenomenon is also perceived when seeing through a transparent medium with internal fractional sub-space metrix ($\mu < 1$), such glass or water.

## Sub-space G graphic representation

The purpose of this geometric analysis is to project motions in sub-space G on a reference flat space, in order to be aware of the distortions suffered by space and time when a remote observer sees motions on a plane perpendicular to his line of sight.

This can be done by projecting 'ds' on a flat horizontal plane, the length 'dr' as measured in the curved sub-space. **Figure XX** represents the phenomenon. The metrix '$\mu$' is the relationship (ds / dr) which geometrically corresponds to cos $\Gamma$, and the apparent velocity ($v_{ap}$ = ds/dt):

$$ds/dr = \cos \Gamma \tag{10-2}$$

$$ds = v_{ap}\, dt \tag{10-3}$$

By definition:
$$\mu = T / T^* = d\tau/dt \tag{10-4}$$

$$\mu = ds/dr \tag{10-5}$$

$$\mu = \cot \theta \tag{10-6}$$

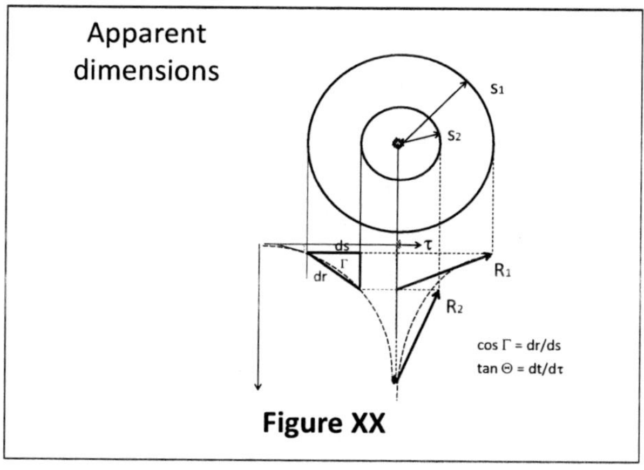


**Figure XX**

168

The metrix is precisely the ratio between travel-times 'd$\tau$' in a sub-space and the absolute-time interval 'dt'; this relation corresponds to (cot $\theta$).

$$\Gamma = \text{arc cos (cot } \theta) \qquad (10\text{-}7)$$

Apparent radii '$s_1$' and '$s_2$' are projections of average distances, so that the perimeter 'p' appears smaller than the actual one (p > 2 $\pi$ s), or ($\pi$ < 3.14).

### Forced motion in a sub-space and magnetic field

An object residing on a sub-space travels on a predefined trajectory only when it is driven by an independently controlled force, other than the one generated by velocity field on which it resides. When an object controls the direction and relative velocity of its trajectory[69], it has to compensate permanently the variations of the central force generated by the emitter while it moves on a non circular path; this phenomenon may be observed in both, electric sub-space and gravitational sub-space.

A receiver travelling on a straight line trajectory cutting the equal metrix circles surrounding any emitter, will perceive an additional acceleration (positive or negative) caused by the increase/decrease of the radial velocity ($\Delta$v) when it changes its distance to the emitter, since the sub-space is a "velocity field" defining a different velocity for every location.

The following analysis finds the magnitude and direction of the force caused by the velocity change (acceleration) when a particle follows a controlled path. In order to visualize the phenomenon, let's refer to a previous example where sub-space is compared to a velocity field generated in a water surface in a container with a drain in its center;

---

69 It could be a self propelled space ship travelling close to a planet, or an electric current flowing inside a copper wire.

we may imagine a radio controlled boat navigating close to the drainage point and following a straight line trajectory (from the perspective of the person who controls it); it is obvious that the boat will be subject to a varying force exerted by the water flowing to the center (gravitational or electric); an additional variable force that is generated by the central speed change when it moves from places where the current is stronger to others with weaker flow; and vice versa; consequently, travelling on a straight path demands a continuous adjustment of the radial force of its motor, compensating permanently the change of the water current speed at that point.

Similarly, any object residing on a sub-space will be subject to radial central forces, which are inversely proportional to the square of the distance to the emitter 'r', this force is generated by the sub-space change of speed 'v' (acceleration) when it moves and therefore changes the speed 'v'; this is because the "velocity field" <u>is a space that moves</u>, inducing a change on its radial position and placing it on a faster velocity position, closer to the emitter. When the sub-space is considered an "electric force field", the force change with radial position becomes obvious.

An object that remains static in a sub-space, not changing its radial distance, must be forced moving away from the emitter at the same radial velocity than the one of the velocity field at that point; this argument shows how wrong is our concept of space and movement; the whole material universe resides on a composite-space which moves permanently; it is like a boat navigating on sea currents. One static object in an abstract flat space build by our mind means that the object must support a force pushing it for getting away from the emitter. Let's analyze what happens when an object moves on a controlled trajectory in a sub-space.

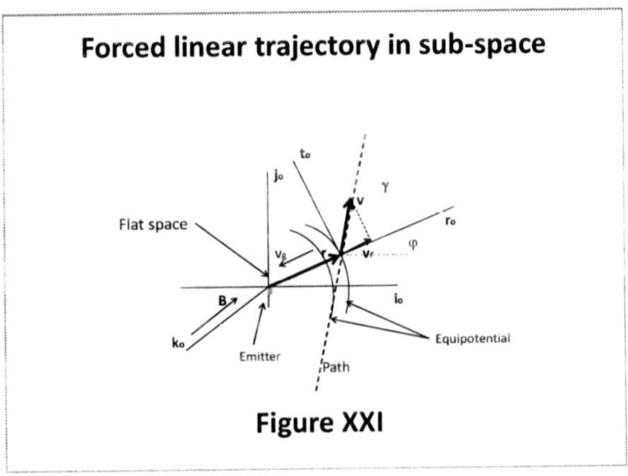



**Figure XXI**

Imagine an object <u>moving on a straight path</u>, perceived as such by an external observer who instinctively locates positions on an imaginary grid on a flat space. Consequently, any object residing on a velocity field which moves the "space", must permanently control a force contrary to the central force at that point; but as it may be also changing its radial position by 'dr', it is subject to an additional force proportional to the sub-space velocity change (acceleration) 'dv', whose rate (dv/dr) is also proportional to its controlled velocity '$v_o$'.

By definition: $\qquad\qquad a = dv/dt \qquad\qquad\qquad\qquad$ (10-8)

And: $\qquad\qquad\qquad v_o = ds/dt \qquad\qquad\qquad\qquad$ (10-9)

Let's assume that a moving particle travels on a rectilinear path at a constant speed ($v_o = ds/dt$) referred to a <u>flat space grid</u>; somehow, the particle self-propulsion is compensating <u>the pull of the sub-space</u> toward the emitter. The acceleration ($a = dv/dt$) affecting the particle when it changes its radial position 'dr' on the electric sub-space is:

$$dv/dt = (ds/dt)\,(dv/dr) \qquad\qquad (10\text{-}10)$$

$$a = v\,(dv/dr) \qquad\qquad (10\text{-}11)$$

Since (ds/dt) is adjusted to be a constant [70] in a rectilinear path as set by the distant observer (projected on a flat space); then, the central acceleration is defined by the variation (dv/dr) of the radial velocity component '$v_r$' from the "velocity field." Previously it has been demonstrated that '$v_r$' in a sub-space is a central vector whose magnitude is inversely proportional to the distance '$r$' ($v_r = k \, / \, r$). Referring to **Figure XXI,** the tangential component '$v_t$' is located on the circle of constant metrix; therefore ($dv_t/dr = 0$); then:

$$dv/dr = dv_r/dr + dv_t/dr = dv_r/dr \qquad (10\text{-}12)$$

and:
$$v_r = k \, / \, r \qquad (10\text{-}13)$$

Deriving:
$$dv_r/dr = - (k \, / \, r^2) \, r^o \qquad (10\text{-}14)$$

Since '$v_r$' is the radial component of the velocity vector '**v**' in the direction given by '$r^o$', then ($v = k \, / \, r$), and:

$$dv_r/dr = - (v \, / \, r) \, r^o \qquad (10\text{-}15)$$

Both, '**v**' and '**r**' are vectors so their quotient is a vector '**X**' conveniently used as an "**intermediate variable**." The resulting vector '**X**' has a magnitude equal to the ratio of magnitudes ($v \, / \, r$) and a perpendicular direction to the plane '**v-r**' in accordance to the right hand rule ($k^o = v^o \times r^o$); then we can define a vector '**X**' as:

$$\mathbf{X} = \mathbf{v} \, / \, \mathbf{r} = (v \, / \, r) \, k^o \qquad (10\text{-}16)$$

$$\mathbf{X} = (v \, / \, r) \, (v^o \times r^o) \qquad (10\text{-}17)$$

As; ($r^o = \mathbf{r} \, / \, r$) and ($\mathbf{v} = v \, v^o$):

$$\mathbf{X} = (\mathbf{v} \times \mathbf{r}) \, / \, r^2 \qquad (10\text{-}18)$$

---

[70] When the object moves at velocities close to the "speed of light", the central attraction slowly vanishes.

It can be observed that in this relation, vector 'X' is similar to the vector known as "magnetic field" 'B' as it is defined in electromagnetic theory. We used this intermediate variable 'X' in order to facilitate the analysis of the phenomena of moving objects on controlled trajectories in a sub-space. The sub-space may be electric or gravitational, as set by an emitter.

The described phenomenon always applies to the interaction between an individual emitter and a single receiver absorbing a photon. The analysis of complex phenomena generated by the dynamic interaction between groups of moving charges is conveniently simplified by defining an abstract entity conventionally known as "magnetic field" 'B', which facilitates the visualization, understanding and mathematical analysis of the phenomena. Consequently, the science of physics must consider the phenomenon of "magnetism", similarly to other entelechies as "space" and "light"; is just a useful abstract associative concept. A useful entelechy is not a substantial entity.

Currently, the magnetic theories study the mutual effects between distant groups of individual entities, assigning to common properties and laws of behavior, but the "magnetism" is not a natural substance, it is just an additional force induced on charges moving in a sub-space.

Considering equations (10-11), (10-13) and (10-14), acceleration is:

$$a = - v \, (v / r) \qquad (10\text{-}19)$$

$$a = - [v \times X] \qquad (10\text{-}20)$$

Or: 
$$a = - [v \times (v \times r) / r^2] \qquad (10\text{-}21)$$

This acceleration, product of the central forces generated by a sub-space E is currently known as "magnetic force or magnetic field"; there is not a conventional name for an existing similar "magneto-gravitational phenomenon" describing the same effect on masses moving in a gravitational sub-space G; it could be the case of the phenomenon experienced by a self-propelled spacecraft travelling at high speed on a controlled trajectory, close to a planet. A similar phenomenon should be observed when a massive rod (or a double star) rotates around its center of gravity, so changing the magnitude of the velocity field on its vicinity, generating variable gravity forces on any receiver close to it. On the contrary, sub-spaces from homogeneous spherical emitters are static; they do not rotate with the emitter[71]. Another case of "forced trajectory" happens when a tying a rope rotating an object; the equation (10-21) is also applicable for calculating its centrifugal acceleration ($a = k\,v^2\,r^\circ\,/\,r$).

Equation (10-21), after applying the appropriate dimensional factors set by the system of units used, is similar to Ampere's law applied to two charges moving at a relative velocity '$v$'; expressed as:

$$F = [(\mu_o\,q^2)/(4\,\pi)][v \times (v \times r)\,/\,r^2] \qquad (10\text{-}22)$$

This equation reconfirms that the formula relating the "intermediate variable" '$X$' with the relative velocity '$v$' and the radial distance '$r$', is similar to the one referred in classic physics as magnetic field '$B$'; consequently, once more is confirmed that the "magnetic field" '$B$' is just a useful "intermediate variable" that helps to understand a phenomenon by simplifying the equations that describe the mutual effect on particles with their sub-spaces in relative motion. Therefore **'B' is not natural substance**. Any equation involving '$B$' should account for this fact; especially when analyzing phenomena involving high relative speeds.

The similarity between equation (10-18) and the Biot-Savart's law, reconfirms once more the accuracy of the herein proposed concept on the phenomenon.

---

[71] The "Coriolis effect", is only applicable when the reference coordinate system rotates with the emitter.

$$\mathbf{B}\ \mathbf{k}^{\circ} = [(\mu_o\, q)/(4\,\pi)]\ (\mathbf{v} \times \mathbf{r})/r^2 \qquad\qquad (10\text{-}23)$$

It is important note also that the well known Maxwell equations are just the result of a pure mathematical analysis on the electric interaction of moving charges, simplified by using as a natural parameter the intermediate variable 'B'.

## Chapter XI

## The magnetic moment

The classic theories fail explaining how a spherical particle behaves as a small magnet when subject to a strong magnetic field generated by intense electric currents; experiments show that it rotates its spinning axis; then, in order to explain the observed behavior, conventional theoretical physics currently assigns mysterious magnetic properties to elementary particles.

The TH shown in the previous chapter how magnetic fields are just a useful entelechy that facilitates the description of the geometric effect on objects moving in a sub-space by explaining the phenomenon with the assistance of the geometric properties of an intermediate mathematical variable 'B', which is not a natural substance but a useful parameter describing the complex interaction between moving charges. In this chapter the TH offers a simple explanation to the observed phenomenon of the seemingly magnetic moment of elementary particles.

Considering that the generic name of "space" may lead to various erroneous results, and that the concept of "sub-space" may refer to a single emitter or a group of them, it is convenient being precise when using those terms.

### Sub-spaces: categories and characteristics

The TH proposes several space categories, each one subject to the particular characteristics and properties; all of them refer to an entelechy similar to the concept of "field" that assigns physical properties to a contingent phenomenon. Once clarified this detail, it will be useful to firstly divide spaces in <u>external spaces</u> and <u>internal spaces</u>.

The external spaces may be divided in several categories, as follows:

A) **Flat space** [72] is a hypothetical abstract concept applied to a homogeneous space with constant metrix. It is used as a convenient geometric frame of reference fitting our human perception of nature. Flat space reference is useful in distant observations of tangential trajectories to the line of sight, or experiments with transparent materials. A natural space, close to be flat is the intergalactic space, with just very few old photons crossing it.

B) **Sub-spaces** are individual circular planar [73] spaces generated by an emitter on its vicinity. All sub-spaces are external abstract entities, normally with fractional metrix. There are two kinds of sub-spaces:

   a. **Gravitational subspace G** surrounding every material body. The mass of the emitter defines its initial metrix.

   b. **Electric sub-space E** surrounds charged particles. It has the same natural metrix than the sub-space G, although sometimes it is convenient to assign a smaller metrix to it, set by an acceleration and the pretended subjacent velocity field, generated by the most appropriate abstract concept of "electric force field."

---

[72] The term flat refers to an isotropic, homogeneous and continuous space with constant metrix equal to one in all locations and in all dimensions; the unit of distance is not a function of position or direction. The intergalactic space is not strictly flat since its metrix is larger than one.

[73] (**) The emitter defines a circular sub-space always coincident with the plane of its ring, as explained in following section.

c. **The magnetic sub-space** does not exist even as a natural contingent entity, but it will be eventually mentioned to facilitate a given explanation.

C) **A composite sub-space X** is the result of adding several individual external sub-spaces generating a complex sub-space related to the geometry of the emitter; it may be the composite sub-space G generated by double stars, or the composite sub-space E generated by a solenoid.

The internal spaces are:

1) **Own-space** describes the metrix of the geometric line that shapes the particle. The own-spaces of particles occupy always only one internal dimension; nevertheless they may be perceived as diverse geometric figures by an external observer:

   a. **The photon's own space** is a <u>straight line</u> whose metrix is defined by the asymmetry of the particle. Since it is the photon generating the external sub-space, sets its metrix by communicating its asymmetry when absorbed by a receiver.

   b. **Ring particles;** like a proton, electron, or neutron have a circular own-space with metrix equal to one, but external observers perceive the ring as with a fractional metrix established by its mass; this is how these particles communicate their mass by emitting compressed photons that form its sub-space.

   c. Other **unstable particles** whose metrix is larger or smaller than one, forming transitory rings eventually exhibiting the properties of charges. They are not analyzed in this study.

2) <u>**Internal sub-space**</u> it is the most abstract concept of spatial environment; refers to the metrix imposed by the

absorption of photons generated by small emitters inside the ring of matter particles. This is the case of a centrical; an electron that has an atomic nucleus proton in the center of its ring.

The metrix is a continuous variable in all external sub-spaces and composite-spaces, because the photon decompression is always continuous.

## Sub-space E particularities

All known stable particles are static structures that may last forever if not disturbed by the absorption of a high energy photon or destroyed by a collision. The induced asymmetry in a free particle when absorbing a low energy photon lasts until it recovers stability by simultaneously emitting a photon; consequently, all free stable particles that build our universe (electrons, and protons) are always charges emitting a Pp[74], so their external sub-space is always a sub-space E. The unstable particles such as a neutron or a centrical (electron around the atomic nucleus), emit an Np that builds a sub-space G.

## Geometry of individual sub-spaces

A material particle is a ring of matter that decodes information when it absorbs a low energy photon, while emitting excess energy as a photon to recover symmetry. From the perspective of an external observer, any photon emitted by that particle always follows a tangential trajectory when leaving the ring; consequently, the sub-space <u>always contains photons travelling on the same plane</u> of the

---

[74] Neutrons do not establish sub-spaces as it will be explained later. A free charge emits a Pp while absorbing an Np because its delicate stability does not wait the finish the absorption process to start emission; it is like a re-assembling engine that restructures the photons. This does not happen with particles in atoms, because they are permanently subject to forces that do not allow them to reach a definitive symmetry.

ring; so the individual sub-space will always be perpendicular to '$\Omega$'. Obviously, when the ring changes its plane of rotation, its sub-space also will change its direction. Considering that already emitted photons are independent entities, any variation on the rings plane, will affect a distant receiver after a TT '$\tau$', so the abstract concept of sub-space describing a contingent entity will have to be represented as a flat surface that may change the direction of its plane; a sub-space generated by an horizontal ring that rotates its plane while emitting continuously in all directions will be a group of rotating circles running away at the "speed of light"; closer points will be on the most recent plane of the ring.

This property is common to any space generated by a material particle, charged or not; the trajectory of an Np or a Pp will always be perpendicular to '$\Omega$'. Therefore any receiver in the vicinity has to reside on the plane of the sub-space to capture a photon; consequently, it is highly improbable that two free particle rings are on the same plane and communicate permanently; this is why any natural phenomena involving distant interaction demands large groups of particles structured as atoms forming a material object as an emitter and as a receiver.

Considering that the size of any emitter is several orders of magnitude smaller that the photons it releases, it is reasonable to assume that any emitter is a dimensionless point in the center of the sub-space's plane it generates.

When the emitter ring changes the direction of '$\Omega$', the receiver will cease to reside on the sub-space plane, so it will not receive its photons. Remember that the particle residing on a sub-space moves at the speed imposed by its "velocity field."

Figure XXII shows an array of detectors circling the ring of the test particle; when this ring rotates on its diameter, '$\Omega$' changes direction, and so does its sub-space. The detector at "a" will cease to receive photons while detector "b" will start receiving them. The experimenter will assume that there is a change in magnetic force equivalent to the one that he will perceive when the particle remains static and the detectors rotate. This interpretation appears more evident when the experiment involves several emitters.

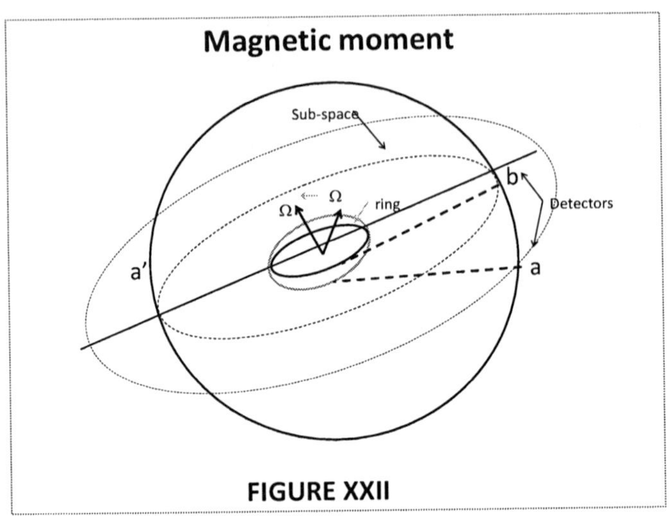



**FIGURE XXII**

An external observer will perceive that when the vector 'Ω' rotates, the receiver at "a" ceases to approach to the emitter; the same effect could be observed if the receiver at "a" moves out of the sub-space plane. It is important to remember that the "space" is defined by arriving photons, so any position has to be referred to the emitter's sub-space; in both cases the particle has changed its vertical distance to the plane of the sub-space. The observed variation in particle's velocity is related to the change in a central force; this phenomenon has been historically interpreted as the effect of a magnetic field on a particle, but in previous chapter it has been explained that this phenomenon is related to the radial forced motion of particles residing in any sub-space.

Consequently, we may conclude that a rotation of vector '$\Omega$' of any particle's ring generates a "magnetic effect" on any receiver. It is absolutely understandable that a scientist who detects a magnetic effect on its sensors will assign magnetic properties to the source, and when the source is an elementary particle will assign magnetic properties to it; furthermore, this erroneous interpretation was absolutely accepted under the current concepts of classical theories.

## Magnetic effect in a composite space

When a free particle resides close to a material object receives signals from the various individual emitters grouped within it, generating a <u>spherical composite sub-space</u>; the resulting force applied to the particle will have the magnitude and direction of the weighted average of the pull exerted by all individual photons arriving to that location in a period of time. When the individual sub-spaces from this composite space rotate, the receiver resting on it perceives a change of magnitude and direction of the central force[75]; this phenomenon is similar to the "magnetic force" that is detected when the receiver changes radial positions within a sub-space, as it has been analyzed in the previous chapter.

Referring to the above Figure XXII; let's suppose that the same figure represents two individual particles, one of them sends photons to "a" and the other to "b"; if the receiver moves from "a" to "b", the result will be equivalent to the one recorded when rotating the object containing the two particles, demonstrating that the analysis of "magnetic forces" in the previous chapter applies also to the effect on the sensor in Figure XXII.

---

[75] This magnetic effect was experimentally observed by Samuel Barnett when investigating the magnetism of objects rotating at high angular speed.

Currently, the intuitive visualization of a "magnetic field" is related to magnet or a solenoid as the artifact used to produce a directionally controlled composite space.

Analyzing the natural phenomenon under the TH perspective, the composite sub-space generated by the current in a helical conductive coil [76] is the integrated total velocity vector created by all charges (moving or static) within the conductor, each one generating its individual sub-space; it is obvious that under such geometrical arrangement the axis of the solenoid will be the symmetric line where the composite effect of all arriving photons will be equivalent to a the radial effect of a sub-space generated by a particle that has the total charge of all emitters.

## Magnetic moment of particles

The classical definition of magnetic moment 'M' is an abstract relation between the torque 'P' produced by a homogeneous magnetic field 'B' on a magnetic sample (solenoid or magnet) given by the vector equation:

$$P = M \times B$$

When this concept is applied to a particle like an electron, the phenomenon is conventionally defined as "intrinsic magnetic moment", so automatically assigning magnetic properties to elementary particles, and therefore erroneously assuming that magnetism is a substantial natural ingredient of matter. Currently there is a common tendency in classical theories to confer magnetic properties to particles; as it has been shown before, this is

---

[76] The production of a current of electrons circulating inside a conductive wire is just a technical trick enabling to the experimenter to place permanently electrons in predetermined positions set by the coil geometry; they could be moving or not; the integrated effect will be the same. The magnetic effect is produced when the receiver moves or when changes the direction of the solenoid axis.

understandable because "magnetic forces" are detected whenever a particle changes the direction of '$\Omega$' as it is subject to a directional torque produced by the absorption of Pp always arriving from a determined direction. Obviously the observed "magnetic effect" will be proportional to the intensity of 'B' which is also proportional to the electric current generating the magnetic field 'B'; as it will be also inversely proportional to the mass of the receiving particle; the Hyle of a small ring rotates faster, and has more mass, having a larger inertia that resist to external forces.

As it is rather difficult to experiment on individual particles, most experiments are performed on large groups of free particles or with objects containing large quantities of particles; in both cases the results are statistically related to the geometry of the field and the degrees of freedom of the particles.

In conclusion, elementary matter particles do not contain a "magnetic substance"[77], neither have a circulating electric currents making them to behave as magnets; it is the sub-space planar shape which produces changes in the force applied on a static receiver. Although this phenomenon is easily detected on electric fields, it also describes the behavior of any external sub-space; including gravitational fields.

Considering that particles are rotating rings of matter, it is quite reasonable to predict that a beam of rings previously oriented on the same plane, will split in two opposed trajectories when subject to a magnetic field; the rings with opposite angular speed '$\Omega$' will move in opposite directions. This is the case of some experiments as the double slit experiment, the Stern-Gerlach experiment and others that

---

[77] Bohr's magneton '$\mu_B$' is an abstract relation originated by comparing the equivalence of the ratio (L / m) between the mechanical moment 'L' of a mass 'm' that produces an angular speed and the ratio between the magnetic moment ($\mu_B$ = P / B) and the charge 'q' producing the particle to spin (S = 1/2); this concept applied to an electron with mass '$m_e$', is ($\mu_B$ / q = $L_d$ / 2 $m_e$) or ($\mu_B$ = q hr / 2 $m_e$).

prove the quantum quality of particles; obviously a rotating ring plane may behave as a binary entity when properly manipulated.

**Precession of particles**

When a torque is applied to a ring rotating at an angular speed '$\Omega$', it changes the direction of the axis of rotation which produces a slow rotation of the '$\Omega$' direction known as gyroscopic precession.

When a free particle residing in a sub-space is subject to the action of the photons arriving from the same direction; every photon applies a force to the rings whose plane is not on the sub-space plane, inducing a change in the direction of '$\Omega$' while it is absorbed; this torque modifies the rotation plane of the ring during the absorption process; this phenomenon repeats every time a new photon is absorbed. This procedure, when properly applied, may be used to align the plane of all rings.

It is very difficult for human nature think freely when analyzing dogmas that supported his own self-confidence and prestige for a lifetime; and worse, it is extremely uncomfortable for anyone to change the goals of investigations in course. History proved unequivocally with Galileo's experience, that this is the initial destiny of all unconventional proposals before recognition.

The TH is not a theoretical guess but a perspective able to explain coherently experimentally observed natural reality; historical observations and experiments prove and will keep proving its solidness.

# Chapter XII

## Nodes

The interaction between material entities is the key process used by nature to communicate the presence and status information among diverse objects. Its mechanics explain how the receiver modifies its characteristics in accordance to the received signal structure, providing a proper instrument to communicate distant entities in a causally related diverse universe; the same mechanism enables particles to link as the building blocks of complex material objects.

The mechanics ruling ring-photon interaction provides an absorption protocol which increases particle energy and changes the structure of the receiver affecting the ring in an inversely proportional way to the photon compression, causing the ring to move in accordance to the effects of gravity and electricity. The same basic mechanics explains the formation of the nucleus and electron centricals [78] in the atoms, and also the refraction/reflection phenomena. The phenomenon of absorption deserves a more detailed analysis than the one presented here; a task for future investigators.

As explained in the introductory chapter, I could not possibly study in full deepness all physics phenomena, including this particular one and others; therefore, this book covers only preliminary concepts and suggestions on the mechanics of nodes formation and the absorption process; moreover, the objective of this study is to explain the basic preliminary concepts, and also suggests possible solutions that may

---

[78] The TH denominates "centrical" to the electron that is part of an atom; therefore, it refers to the ring of Hyle centered on the atomic nucleus; not to the small spherical electrons orbiting the nucleus as described by earlier atomic models, nor the abstract mathematical variable described as a probability function by the quantum mechanics.

lead to find precise mathematical relations describing every particular phenomenon, from the perspective of the TH.

Before studying this subject in detail, it is important to examine and revise some previously presented, but significant concepts.

**External space straight trajectory**

The external space is just the mental image of a stable phenomenon. When a three dimensional volume does not have photons crossing it permanently, the term "space" fades away remaining just its abstract concept that does not exist really as a natural entity. Always, when analyzing the characteristics of "space" we must be aware about its contingency; for instance, based on our knowledge that the universe is a causal entity, we must know the past history of a particular sub-space, in order to increase the prediction probability on its future properties.

It becomes obvious that under the TH perspective, the minimum distance trajectory, known as a straight line, will be the one that follow the photons connecting two positions; therefore photons travelled distance '$\tau$', defines the 'straight line' that is not always straight when compared with a flat space reference; which means the linearity of its function with respect to 't'; so there may be cases where '$\tau$' may be a higher order function of 't'.

Both, human semantics and conventional physic consider "space" as a physical substantial entity; so it is important to define the real meaning of "location" and movement when applying the concept of "space" in the analysis of basic interaction process among two Hyle structures having some part of them occupying simultaneously the same point in an N dimensional field.

Let's analyze initially the macroscopic perspective of the phenomenon; imagine two equal masses as the independent sources of low energy photons; if "A" emits higher energy photons than "B",

<u>an equidistant receiver</u> absorbing alternatively one photon from "A" and one from "B" will be subject to equivalent gravitational attraction from both emitters; so we may conclude that both sub-spaces have similar properties, and that the individual external sub-space metrix is independent of the energy of the photons. This conclusion is coincident with what was demonstrated in previous chapters that the external sub-space metrix depends only on the relative compression of the photons crossing it.

When an object moves through a sub-space G, it will follow a conical trajectory ruled by Newton's law; but when the receiver trips through a composite space as the one generated by two emitters like the one described above, its trajectory will not be conical anymore, proving that the geodesic lines that define the straight trajectory always depend on the geometric location of the sources that generate a particular "space."

It is also important to remember that the consolidated effect of multiple signals arriving to a receiver from various sources will be a statistical process whose results will depend on the relative quantity received from each one; so, the relative "brightness" of each source, defines its relative influence on the receiver. Let's suppose that "A" is 100 times brighter than "B"; this means that the receiver will be subject to the action of 100 photons arriving from "A" for each photon that comes from "B"; so the concept of sub-space always refers to a non Newtonian statistical process because the composite metrix implicitly depends on the relative quantity of photons received; consequently, an object not emitting photons, as a hypothetical black hole, will not communicate its presence to any object; it will be just a mental entity that does not exist in our physical universe; consequently, it is reasonable to assume that the quantity of photons crossing a location, <u>set the prevalent direction of the "inertial path"</u> (also an abstract concept) of the "composite space"[79].

## Photon straight path

The concept of distance is: the lapse of "travel time" that a point of a Hyle structure spends travelling on a straight path from one point to another in an environment (space) where the permanent presence of Hyle entities passing through it define its metrix and curvature.

Although normally a "space" in nature refers to a central field generated by a massive emitter, theoretically could exist an isotropic space generated by many gravitational equivalent sub-spaces generated by equally compressed photons arriving from all directions, in this case the straight path that follows a photon can be a curve because the physical meaning of "straightness" is closely related to the geodesics generated by the prevalent path of the signals crossing it; a similar case will be given when there is a complete absence of signals as in the beginning of time. In both cases, the straight line is also absolutely undefined, so the Hyle structures could follow any path: straight, curve or completely closed as perceived by an external observer; consequently, all those paths will statistically exist in such environmental conditions.

Other extreme situation appears when there is always a Hyle structure crossing a given position in a sub-space; this is the case of matter particles. The photon that is travelling a straight path from an external sub-space and arrives to the particle own-space, we may assume that the photon will keep traveling on a straight trajectory, which in the own space of the particle is defined by the circle of the ring; because

---

[79] To clarify the explanation we may compare the receiver to a boat floating on an oil spillage being drained out upstream on a river; the current of the river will pull the boat more than the oil layer being pumped out.

the importance of the particle Hyle structure define its geometrical appearance[80].

It is important to note that all particles (photons and matter) are one-dimensional structures that occupy only one spatial dimension with zero thickness, so a photon may cross inside or outside from any particle ring, no matter how small is it; the probability of a photon to touch a particle is always very small, but this probability increases when the ring rotates on its diameter while a photon passes inside it; since most photons are very long structures in comparison to any ring diameter (if we represent an electron as a circle with a diameter of one meter; a proton will have a diameter of ½ mm, and a photon will be 200 Kilometers long).

A photon may also collide with a particle ring under special circumstances and then rebound; as happens with mirrors[81]; obviously this is a statistical process, which explains how diverse materials have a given color; how they absorb/reflect light or heat, etc...

## Structure of photons in a particle ring

The metrix of the own space from particles is defined by the compression of its Hyle; its trajectory is a straight line independently from any external sub-space because there is always Hyle travelling on its own-space. Consequently, the "own space" characteristics are permanent and applicable to any point in the perimeter of the ring because all of it is permanently occupied by Hyle; when one of the vectors '$H_1$' forming it moves from a position to another on the perimeter of the ring, there will always be a new vector '$H_1$' occupying that location immediately. Therefore, the "own-space" of a ring is not

---

[80] Note that the quantity of signals represents spatial energy density; this is why the own space of a ring prevails over the small spatial energy density of a photon.

[81] The reflection phenomena deserve to be deeply analyzed because it may offer a method to decrease the influence of gravity and electric fields on massive receivers.

a statistical entity; it permanently exists, even when the particle does not receive or send signals.

This stable object (the ring and its own circular space) may be placed inside a sub-space, which is a probabilistic entity that may cease to exist when the surrounding emitters stop emitting signals. This argument shows clearly that matter may exist forever as a static Hyle structure, independently from any external environment when its symmetry is not disturbed by some arriving signals, but it will cease to be part of our universe when not receiving signals or subsequently emitting its own signals containing information about its characteristics.

The absence of signals is not the case in our observable universe; even the most distant galaxies keep sending signals Only when all matter becomes extremely cold, there will not be signal interchange, so gravity will fade away, and atoms will dissolve because protons will not be able to attract electrons; therefore, the final fate of universe is to arrive to a condition where permanent communication is absent.

The "arrival" of a photon to an isolated entity is the way that nature expresses that something out there sent signals that travelled some TT before arriving to it. As soon as a photon touches the particle, there are two alternative behaviors: it rebounds or it keeps travelling on a straight path defined by the particle's Hyle and it follows a circular stable path[82]. When this happens, its external TT '$\tau$' does not increase anymore since the straight line will be seen by any external observer as a very small circle; so it can be properly affirmed that signal has arrived to destination.

---

[82] The explanation simplifies the phenomenon by using the term "space" as being a substantial entity, which is not strictly proper. The objective in using this description is to simplify the analysis of a mechanical process similar to a pre-compressed rope winding-up in a pulley.

The phenomenon may be described as a photon winding-up on the receiving ring [83]; but taking into account that it refers to the coexistence of two independent entities circling at different pace: the Hyle of the ring is a symmetric structure that sets its own-space as the clock face showing the absolute universal time on the perimeter, so its '$\tau$' is always equal to 't'; on the contrary, the Hyle of a compressed photon is an asymmetric structure circling at a lower rate because 'T' is smaller than 'T*'.

## The nodes

The TH postulates that Hyle is the only substance constituting all particles as independent entities formed by rotating vector structures that travel a "distance" 'T' in one specific TT dimension while rotating a whole revolution in a time 'T*'. The geometry of the particle defines the direction of the vector function on each position of the arrangement; so, the direction that each one of its vectors is pointing to a different direction in a 3D TT space[84]. Normally, the length 'T' of a particle, should be the same than its revolution period 'T*', in AT units; this is the way it <u>defines the unit of distance</u> in the TT spatial dimension it travels.

The Hyle vector function may be also represented as an <u>energy density vector 'H' packed in the TT dimensions whose amplitude is binary</u>; always one within the spatial limits of a particle and zero outside it, so establishing its length as the TT segment occupied by it.

We may represent Hyle vectors <u>as a train of individual vectors</u> symbolizing that there is always a single vector pointing to some direction in that spatial position, its magnitude cannot be algebraically

---

[83] A detailed analysis of this second alternative will be developed, but it can be also applied when the photon rebounds, because by curving at contact point, it will have similar behavior.

[84] The Hyle structure always travels in one TT dimension while its vectors, depending of the type of structure, rotate on a plane, known as its "disc", pointing to any of the three TT dimensions.

added because <u>it cannot exist twice,</u> because the vector 'H' defines existence; the vectors 'H' cannot be added as ordinary vectors because may result an absurd; that the entity it defines exists fractionally, or has more than one existence.

Let's analyze in more detail this phenomenon. When two particles touch, it means that they share momentarily the same spatial location, so their vectors in that position and in that precise moment will add, but the result of this addition must be binary, one or zero because vector 'H' indicates "existence." When both vectors point to the same direction, the result cannot be 2, because it would mean that there is an Hyle structure (matter or energy) that has a segment existing twice in that point; when its angle is other than $180^\circ$, the result would be that the structure exists fractionally in that point; is also an absurd accepting the concept of a partial existence. In these two cases the particles collide resulting that both will preserve their individuality and independence.

There is a third case when their vectors oppose; in this case the Hyle of both particles ceases to exist in that point. When the affected structure is a ring, a part of its perimeter disappears; and if it is a photon, this sector it will cease to exist; <u>it will be absorbed by the ring</u>.

This annihilation process has two consequences on a ring:

> - It will cut a section of the ring perimeter[85]; but as the straight path in the own-space of the particle is a circle, the Hyle will keep travelling circularly, <u>while the discontinuity is repaired</u> by action of the internal elasticity.

---

[85] This study analyzes node phenomena involving two rings or a ring and a photon. A collision between photons is quite improbable, but when it happens, they will alter their paths. It is also extremely improbable for a photon to absorb another photon.

- In the AT dimension it will <u>immediately</u> shorten the period 'T*', increasing the angular velocity of the ring '$\omega$' and therefore particle's energy.

In conventional language the absorption process is an energy exchange that increases the energy of the receiver by decreasing its period '$T_1$*' in the amount of the incremental energy absorbed from the photon, while a <u>tensional stress</u> and <u>a torsion</u> is applied to the particle structure.

From a spatial perspective, the permanent presence of a node among two similar particles touching each other means that some part of both particles share a spatial location. This point is an empty space because in the case of similar particles the inexistence of Hyle at a point means <u>the permanent presence of two opposed Hyle vectors</u>. Two similar entities, sharing one or two nodes, keep their independence, so the node represents a common spatial position bonding both until one of them changes its '$\omega$' by drastically changing its diameter.

When one of the particles ceases to exist, as in the case of a photon, which disappears after being absorbed by the ring, the particle perimeter is accelerated, and recovers its symmetry by filling the empty spaces by rearranging the angular position of its Hyle by decreasing the ring diameter. The other alternative is that the generated shear is so high, that it fills the nodes in the perimeter by releasing one of the two opposite vectors 'H' that are building the node.

## Photon-Particle interaction

A photon builds a node by winding-up on a ring, as a rope whose end was fixed to the edge; let's suppose that the Hyle from both entities do not mix, so the Hyle of the photon travels parallel to the perimeter of the particle applying a force and a torque to the lengthy photon structure, which is proportional to its compression.

Photons are vector structures rotating continuously that contain mass (or energy) in accordance to Einstein-Planck relation. Therefore, this structure may be "digitized" [86] so it could be represented by a chain of similar vectors pointing to a slightly different direction along the photon's length. The analysis of the interaction process is facilitated by describing a photon as formed by 'n' vectors, each one containing a fraction '$\varphi$' of its total energy; let's set the distance between these vectors to be equal to the perimeter of the ring, so ($\varphi = T_1^* / T_o^*$) where '$T_o^*$' is the photon period and '$T_1^*$' the ring period (in the case of a visible light photon and an electron this ratio is smaller than $10^{-5}$).

Initially, the receiver is always a symmetric particle, so the geometric position of the vectors from a photon, winding-up circularly on a symmetric ring depends on its compression; on the contrary, the ring vectors occupy a "flat space" defined by its symmetric structure. A symmetric photon will overlap sequentially its first vector '$H_o$' with every new vector '$H_o$' in every new turn, both <u>pointing to the same direction</u>.

The length '$T_o$' of a compressed photon with metrix ($\mu < 1$) is smaller than the period '$T_o^*$' ($T_o = \mu T_o^*$); so, if not stretched, it will have its energy vectors '$H_o$' spaced at a slightly smaller distance than the perimeter of the ring since ($T_1 = T_1^* = \varphi \mu T_o^*$) and will point to a slightly different direction, as illustrated in the next Figure XXIII.

---

[86] Although the interaction between two continuous functions should be continuous, it is absolutely valid (as demonstrated in classical differential calculus) to study the process by analyzing its incremental behavior to facilitate its comprehension. This approach, although mathematically valid, is not strictly valid in physics because energy is proportional to angular velocity, independent of distances within a particle; one cannot segment "an angular speed."

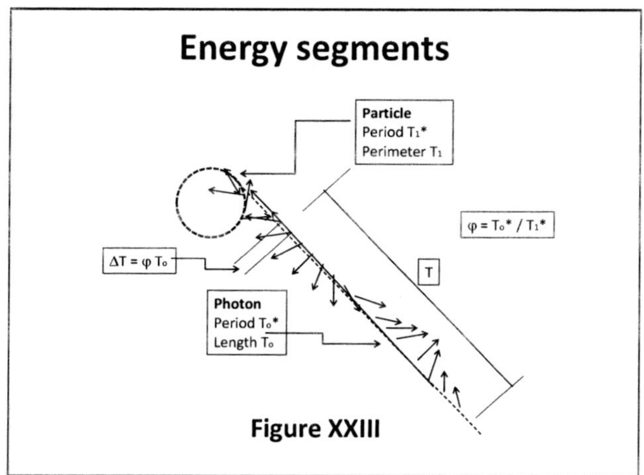



**Figure XXIII**

Consequently, it is reasonable to assume that: when all the Hyle from an asymmetric photon is wrapped circularly on a ring, the photon could be represented as a new circular entity [87] containing the total photon energy as a series of individual equally spaced energy vectors concentrated in a sector of the ring, each one representing the fraction '$\varphi$' of the photon's energy.

Under the same perspective, the proposed model represents a particle as formed by a large quantity of equally spaced vectors '$H_1$' pointing permanently to different but permanent direction at every perimeter position; similar to the marks on a clock's face; this is because a symmetric ring is a static entity where all '$H_1$' vectors permanently point to the same direction.

Any Hyle vector in the present analysis represents an amount of energy arbitrarily defined as the unit of energy '$\varepsilon_x$' for both interacting

---

[87] By applying the proposed model, a circular photon <u>does not represent a natural continuous vector function</u> since it is rotating at the same angular speed of the ring but it does not have the same energy given by the Einstein-Planck relation. So we may conclude that just few vectors could be circling simultaneously before disappearing one by one in a node. The real behavior must be analyzed by studying the continuous energy buildup so a new node is produced before a new '$H_0$' enters the ring own-space.

particles, the photon and the particle; consequently all vectors will have an amplitude equal to one, so a Hyle structure will contain 'n' vectors 'H' in proportion to its total energy, and they will be separated by a distance (T / n) along the total length 'T' of the particle.

A second important factor to be considered is that an undisturbed ring is always a symmetric entity with ($T_1 = T_1{}^*$), so it looks static to an external observer because its temporal phase is the same for every angular position; consequently the series of equally spaced vectors from the asymmetric photon entering the own-space of the particle, will travel on the perimeter at a slower angular speed[88].

A third consideration to be taken into account is that the absorption process is sequential, so the inner ring will form nodes one after another while the absorption process lasts, as explained in the next subtitle.

### Absorption model of "Energy Segments"

The roots of the absorption phenomenon are related to the mutual effect on the curvature of two independent Hyle structures sharing a location in a four-dimensional time environment. The precise analysis of this difficult subject falls out of the scope of the current publication; nevertheless it deserves to be deeply analyzed by future investigators by using the most appropriate mathematical tools for the task.

Considering that the direction of the each Hyle vector of a particle is the most basic parameter that defines the presence of the segment of a particle in a 4D time environment, we will propose some convenient premises to explain the interaction among two particles; these

---

[88] We may imagine observing two concentric rings; the external ring resembling the particle will look like a gear whose teeth always point to the center, and the inner ring as a circular helix rotating counterclockwise because it is compressed; its length is smaller than the perimeter of the circle.

precepts are not strictly valid from the physics perspective [89] although will be geometrically useful in the analysis of the phenomenon. We will assume that a particle with a given length 'T' contains energy '$\varepsilon$' can be segmented in small longitudinal segments each one containing the same portion of energy and represent the segments as energy vectors that point in the direction of the Hyle 'H' vector it that spot of the particle[90]. The same analysis could apply the concept of "dimensional existence" of the Hyle vector instead of using small energy vectors, but it may be more easy for the reader to think is terms of energy portions than in abstract vectors. Throughout the analysis that follows, it must be kept in mind that the essential concept of energy is related to rotation and curvature in a 4D time environment, more than to location and distance.

When a photon enters to the circular space of the particle winding-up its linear structure on the perimeter of the ring, it slightly changes the direction of its vector 'H' with every turn, so it is convenient to divide its total energy in "$n_o$" segments, each one containing a fraction of photon's energy. It will be defined that the vector representing that segment of energy will have the same direction than the Hyle vector has in that position of the photon arrangement.

Let's define the energies of the photon and from the particle, as given by the Einstein-Planck relation:

The particle energy is:
$$\varepsilon_1 = h \, / \, T_1{}^* \qquad\qquad (12\text{-}1)$$

---

[89] This assumption is strictly incorrect since the particle energy is defined by its period, which eventually may be equal to the particle length in symmetric particles. In other words, the energy is a property given by its Hyle rate of rotation from which the length is just a secondary contingent outcome; by dividing a particle in two, one must break it in two particles rotating at half angular speed by doubling the period; not by cutting it in half.

[90] The contradictory concept applied in this analysis is to assign a length to each one of energy segments when energy is a parameter inversely proportional to length. The consequence is a positional indeterminacy of nodes.

The photon energy is: $\varepsilon_o = h / T_o{}^*$ (12-2)

When dividing the period '$T_o{}^*$' of the photon in '$n_o$' segment each one lasting an AT interval equivalent to a complete ring revolution '$T_1{}^*$', the photon number of segments will be ($n_o = T_o{}^* / T_1{}^*$), each one of them represented by a vector '$H_o$' that will contain an "energy packet" ($\varepsilon_x = \varepsilon_o / n_o$). Therefore, the energy '$\varepsilon_1$' of the ring will embody '$n_1$' vectors '$H_1$' of the same energy '$\varepsilon_x$', so ($n_1 = \varepsilon_1 / \varepsilon_x$). Considering that the energy ($\varepsilon_x$) is a minuscule fraction ($\varphi = T_1{}^* / T^*$) of total photon energy '$\varepsilon_o$'; so, we may suppose that '$n_o$' and '$n_1$' are whole numbers[91]:

$$n_o = \varepsilon_o / \varepsilon_x = T_o{}^* / T_1{}^* \qquad (12-3)$$

$$n_1 = \varepsilon_1 / \varepsilon_x = (T_o{}^* / T_1{}^*)^2 \qquad (12-4)$$

After some revolutions, the first vector '$H_o$' entering to the circular space of the rotating ring, will find an opposite vector '$H_1$', producing the first node; the process will repeat for the rest of the vectors until the total photon energy is completely absorbed by the ring.

The following paragraph explains some preliminary concepts about the complex geometry of the absorption process and its consequences. The detailed equations governing this important phenomenon deserve to be explored in full detail by future investigators.

### Basic protocol of node formation

An imaginary and isolated vector 'H' travelling into a circular space does not change direction because it does not belong to a rotating structure, but all Hyle vectors in nature <u>are always part of a continuous rotating structure;</u> this structure moves while simultaneously rotates as a whole, so each vector representing a

---

[91] For an electron ring and a visible light photon, ($n_o = 10^5$) and ($n_1 = 10^{10}$). These quantities are 2000 times larger for a proton or for a neutron.

portion of energy cannot be isolated from the others that are in the structure; it belongs to an elastic substance that resists changes, so there is always some inertia and resistance to overcome when some part of this structure changes direction; this concept is applicable to any particle, including a very low energy photon.

When two vectors of similar energy, share the same spatial position simultaneously there can be two possible effects: they form a node, or they collide (they never add up or mix); so that both structures get distorted while run separately in accordance to the laws ruling elastic collision. When one of the colliding particles is a photon with an almost negligible mass compared to the ring mass, it will rebounds, in a similar way that rebounds on water a fishing line without a hook, twisting its trajectory in accordance with the very well known reflection laws, so the path curvature at contact point all '$H_o$' vectors bend their trajectory and pull the structure in the new direction.

When the first portion of the photon, a continuous entity, touches the particle, the first vector '$H_o$' containing a fraction of energy '$\varepsilon_x$'; if hooked, enters the ring own space; this vector rotates on the plane of its disc which always maintains the same angle with the direction of photon path, even when it enters to the circular own space of the particle; so '$\omega$' has the same angle with the tangent of the ring than the one it had with the straight trajectory it traveled in the external space before entering to particle own-space.

Since the photon metrix is smaller than one, the vector '$H_o$' rotates faster on its own disc than the Hyle of the ring does when both structures travel the same distance on a flat space (circular or linear). Each one of the '$n_1$' vectors of the ring always points in the same direction along its position on the circle[92], so the ring position

---

[92] For an outside observer the particle is perceived as a circle so the vectors '$H1$' appear always pointing in the same direction because they advance the same angle on the circle as they rotate the same angle on its disc.

represents <u>the geometric (time) reference</u> strictly proportional to the AT. The first node appears when the initial vector '$H_o$' intercepts an opposite vector '$H_1$' somewhere on the ring perimeter.

The asymmetry induced by every new node in the ring causes that all '$H_1$' vectors be no longer static; all them start moving faster because '$\omega$' of the Hyle increases while the magnitude of the angular velocity of the ring '$\Omega$' remains; this is because every new node decreases '$T_1{}^*$' instantaneously while '$T_1$' decreases very slowly because the ring takes some time for recovering symmetry; this means that at the moment of absorption, the ring metrix ($T_1$ / $T_1{}^*$) increases and becomes larger than one, which increases ring's tangential velocity while recovering symmetry. On the contrary, the compressed photon has a metrix lower than one, so this Hyle moves slower than the Hyle of the particle, while both of them travel in the own-space of the ring.

It is important note that the effect of nodes increasing ring metrix is irrelevant during the absorption period, because the energy increase caused by a typical node is a million times smaller than the total energy of the particle. The total duration of the absorption process depends on two factors: the energy and the asymmetry of the photon; but in any case, the ring rotates hundreds of thousands of times before absorbing completely a photon.

## Detailed absorption procedure

To analyze the basic mechanics of the absorption process it will be convenient establish a nomenclature which facilitates the description of the parameters involved in the phenomenon. The Figure XXIV shows a vector '$H_o$' rotating on a plane which we will call "disc $H_o$" with an angular velocity '$\omega_o$' (period 'To*'). This photon with energy '$\varepsilon_o$' is compressed (it has a metrix $\mu < 1$) with a compression 'z', so it has a length ($T_o < T_o{}^*$). Its whole structure always travels on a straight path in the external sub-space as it does in the circle (also straight) of the own-space of the particle.

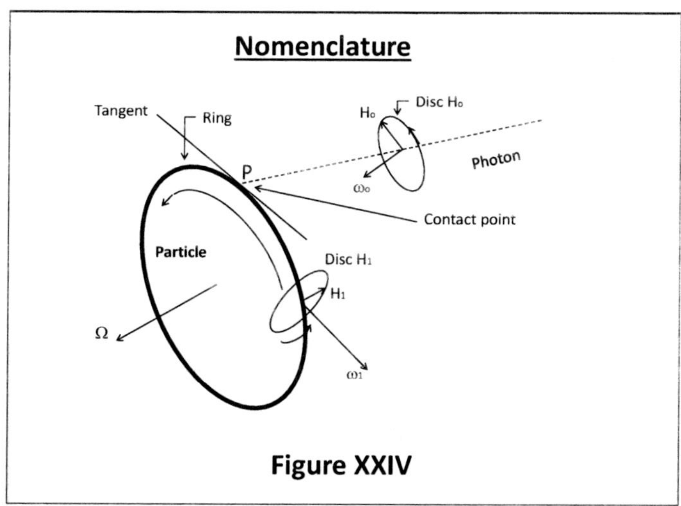



**Figure XXIV**

When a compressed asynchronous photon enters the own space of the ring, like a thread on a spool, the trajectory of that part of the photon follows the circular ring path, but as it is asymmetric, an external observer will perceive that the vector train '$H_o$' entering to the ring appears receding, until they build a node when finding an opposite vector "$H_1$" from the ring. The detailed process is much more complex than this simple description; a more detailed, nevertheless incomplete, analysis follows.

The process of photon absorption will be described based in the assumption that it enters to the own space of the ring; a very similar analysis could be applied assuming that the photon rebounds and curves its trajectory to the inner part of the ring at the contact point; therefore, the following analysis is applicable to both models:

-   The vectors '$H_1$' of the receiving particle constitute a ring-shaped structure which rotates at an angular velocity '$\Omega$'. The "disc $H_1$" is the plane on which every Hyle vector '$H_1$' rotates at an angular velocity '$\omega_1$', (period '$T_1{}^*$'). The ring has a perimeter '$T_1$' which is numerically equal to '$T_1{}^*$'

because a stable receiver, by definition is always a symmetric structure.

- When the two structures meet at a point "P", their respective discs intersect on a line "XXp" whose direction generally does not coincide with the plane of the ring. Both vectors '$H_0^P$' and also the vectors '$H_1^P$', have a different angle with "XXp." Normally the angle between both vectors will not be precisely 180°, so they will not eliminate each other in order to form a node.

- Two rotating vectors '$H_0$' and '$H_1$' which momentarily are the same spatial position "Y" on the periphery of a ring will not always be oriented in the same direction; their discs intersect at a common line '$XXy$' where each vector will have a different angle with this line; '$\phi_0^y$' for '$H_0^y$', and '$\phi_1^y$' for '$H_1^y$', as shown in Figure XXV.

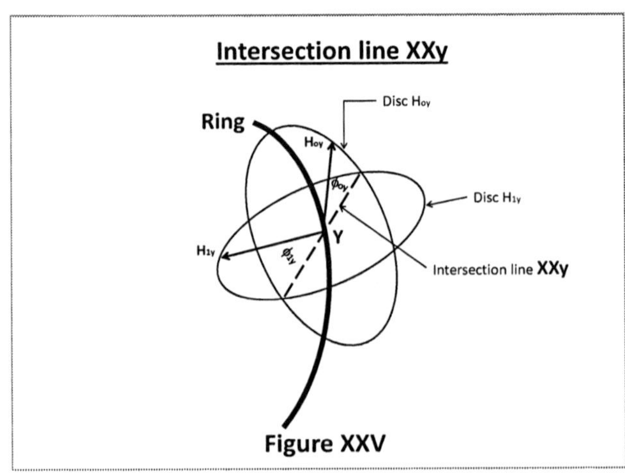


**Figure XXV**

- The duration of the interaction between these two vectors is ($T_1 / n_1$) because they occupy an angle ($2\pi / n_1$) on the ring perimeter; although both vectors contain the same energy, the wideness of '$H_0$' is along most of the perimeter, while '$H_1$' is precisely located in a point of the perimeter. The duration of their interaction <u>is</u>

independent of the angular velocities 'ω' of their structures; this important fact shows that the construction process of a node <u>is independent of the direction of the individual paths from the photon or the direction of rotation of the ring;</u> the phenomenon is limited to the momentary interaction of two vectors pointing to different or to the same direction.

- A node is formed only when three conditions are met: a) the vectors of both particles '$H_1^y$' and '$H_o^y$' meet at the same position "Y" on the perimeter of the ring, b) <u>both are on their discs intersection line "XXy"</u>, and c) they point to opposite direction.

- The first photon vector '$H_o^p$' arriving at contact point "P" normally will not form a node with a ring vector '$H_1^p$' because it will be extremely improbable that both coincide and point in opposite directions simultaneously. The vector '$H_o^p$' is part of an asymmetric structure, so gradually decreases its angle with the intersection line "XXy" as this line travels on a point "Y" of the ring perimeter rotating $360^o$ in a slightly longer time that '$T_1^*$'. The photon disc maintains its angle with the tangent of the ring as it enters the circular space of the particle because it keeps travelling on a straight line. The particle vectors '$H_1$' and their respective discs are static, so each one of them keep invariable their relative direction with the tangent. In conclusion, the discs of the photon travel around the ring perimeter in a time '$T_1^*$' keeping the same angle with the tangent as the initial one at contact point "P."

- The angle '$\phi_o$' between the first photon vector '$H_o^p$' and the intersection line "XXp" at contact point "P" depends on several factors: a) the direction of photon arrival, b) the type of structure, c) the ring plane, and other parameters; so this angle could be anyone. Observing the ring, the angle '$\phi_1$' of vector '$H_1^P$' with "XXp" will not always be zero; therefore, generally it is very unlikely having a node at the contact point. The normal process will be that the photon enters the ring own-space or rebounds; depending on a range of the angles between the two 'H' vectors at "P." Considering that '$H_o$' vectors change directions very slowly, they are very wide compared to '$H_1$' vectors that are extremely narrow; this fact makes it easier for the first vector '$H_o$' builds a node because its angular phase is imprecise, we may consider that it occupies a large portion of the perimeter.

- The first vector '$H_o^P$' that enters to the own-space of the ring, travels over a point "Y" on the circular path while the ring itself rotates at an angular speed '$\Omega$'; the angular speed of "Y" is slightly smaller than '$\Omega$' due to the photon compression. For the same reason, the vector '$H_o^P$' rotates on its own disc at a slightly larger angular speed than '$\Omega$', so decreases its angle with "XXy" in every revolution.

- The angular velocity ($\omega_o = 2\pi / T_o{*}$) of '$H_o$' is independent of the asymmetry of the photon, but the wave number ($2\pi / T_o$) depends on the compression '$z$' of the photon, so the distance between consecutive '$H_o$' vectors are a function of the compression.

- The line "XXy" refers to the intersection between "two discs at any point "Y" on the ring" and " XXp" refers to the intersection of the disc from the same vector '$H_o$' at contact point "P." Although the angle '$\phi_o$' changes slightly during every revolution, and the vectors '$H_1$' are not static when the ring contains nodes, in order to facilitate the analysis we will consider that both remain constant while a vector "XXy" completes a revolution; therefore we will assume that before starting a new revolution every vector

'$H_o$' suddenly gets closer to the intersection line "XXp", until angle '$\phi_o$' is reduced to zero.

- After some time '$\phi o$' becomes zero, so the vector '$H_o{}^a$' overlaps "XXy." Then, the intersection line 'XXy' sweeps all the '$H_1$' vectors on the ring that sequentially point to all $360°$ directions on their disc at every location on the ring perimeter; therefore "XXy" will find a vector '$H_1{}^a$' at position "A" pointing to the opposite direction of '$H_o{}^a$', so both disappear, establishing the first node.

- The next FIGURE shows how <u>the disc '$H_1$'</u> along the ring maintains an angle '$\varphi$' with the tangent as it rotates on the ring at an angular velocity '$\Omega$'. It also shows how vector '$H_1$' rotates on its disc at an angular velocity '$\omega_1$', which changes its angular position along the perimeter, while maintaining the same direction when it is at the same position on the ring, because it is part of a symmetric structure.

- The disc '$H_o$', travels along the ring at an angular velocity smaller than '$\Omega$', but the intersection line "XX" maintains its initial angle with the tangent of the ring (in position XXa or XXb). This is because the plane of the tangential angle of the <u>'Ho' disc</u> remains constant.

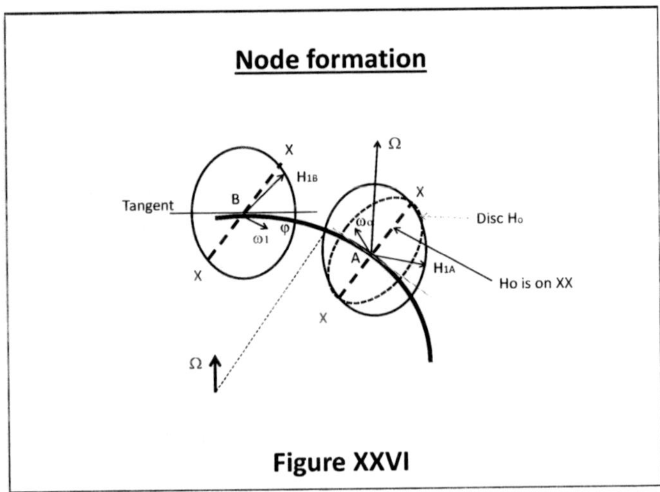



- Let's suppose that the first vector '$H_o^{P}$' finally arrives to the contact point "P" after many revolutions on the ring and <u>overlaps the line "XXp"</u>; at this moment the next vector '$H_o^{2}$' has an angle with the intersection line "XXp" equal to $(2 \pi / n_o)$[93].

---

[93] The above description suggests that the first photon '$H_o^{P}$' is "absorbed by the particle own-space; but one may get similar result considering that a photon that collides with the ring bends its trajectory producing a directional change in its first disc that may suffice to build a node while the ring completes hundreds of revolutions.

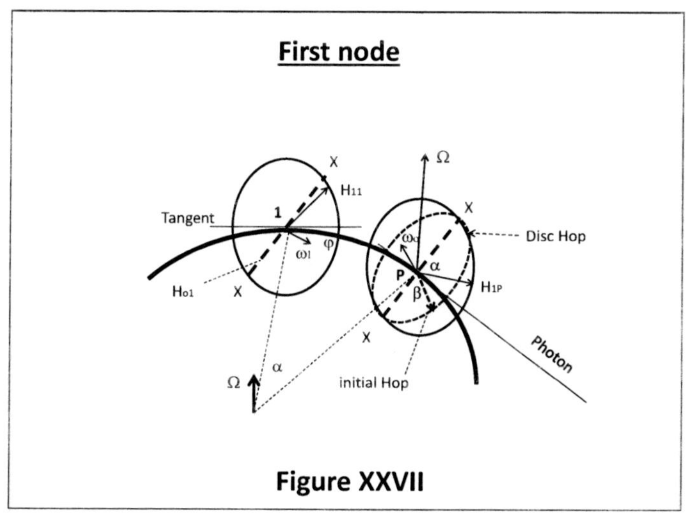



- The angle between '$H_o^2$' and "XXp" that was $(2\pi / n_o)$ when the '$H_o^1$' established the first node at "1", will be zero after '$T_1{}^*$', because '$H_o^2$' position was defined under this specific condition, established when dividing photon length '$T_o$' in '$n_o$' equal energy sections <u>spaced at equal AT intervals '$T_1{}^*$'</u>. This demonstrates that the second vector '$H_o$' is at "P" will overlap "XXp" independently from '$\omega_o$' direction.

- In any case, the first node position "1" on the ring secures the front end of the photon on the ring structure, which <u>pulls the whole photon structure to rotate around the energetic ring</u>, winding it as a slim string around a wheel. But it must be remembered that a compressed photon with metrix ($\mu < 1$) increases its TT at a slower rate than the Hyle of the ring, so it will be subject to a force pulling its structure in a similar way than an elastic band is stressed when winding-up on a rim; this pull produces a slight change '$\alpha$' on the '$H_o^2$' disc plane direction at "P" as explained bellow.

- Before the first node nulls '$H_o^{1}$' at point "1", both structures were independent, so the photon was not tensioned. When the first node appears, the photon linear structure will be tensioned by a force ($F = k\,\xi$), so the next vector '$H_o^{2}$' will be forced to <u>rotate on the '$H_1$' disc</u> while the photon moves on the ring perimeter as explained in the next paragraph; so, after the node "1" completes a revolution vector '$H_o^{2}$' will be close to position "1" pointing to a slightly different direction than '$H_o^{1}$' had, when formed the first node.

- The structure of any particle could be compared to a line in the 3D spatial environment but this line has an additional property; it exist because it is formed by a vector structure <u>that points to different directions in each point of its structure</u>; this characteristic could be visualized as a twisted band. In accordance to this description, the particle's ring is a wheel with a warped "surface" on which a very long and slightly twisted elastic tape representing the photon, rests tensioned on the "surface" of the ring; but an additional condition must be observed, that this "warped surface" repels the band representing the photon because their '$H$' vectors do not point to the opposite direction.

- Consequently, the tension produces a small phase shift '$\alpha$' of vector '$H_o^{2}$', which is larger when the compression '$\xi$' is large and vice versa[94], this effect is caused because the ring forces to the photon structure share the geometric shape of its perimeter; this is the same than saying that '$H_o^{2}$' tends rotate with the Hyle of the particle as the result

---

[94] Considering that the photon is a continuous entity, this representation for the analysis is imprecise because the real particle is not a series of isolated vectors but an unbroken function that loses a portion of its length. This characteristic makes it possible to exert a pull because the longitudinal integrity of the particle exists until it is completely absorbed.

of the tension generated by photon elasticity because of the length differences among the symmetric ring and the compressed photon that additionally is being repelled. By this mean, each new node will be produced on the ring in the neighborhood of the first node, closer when the photon compression is higher, so the angle 'α' between nodes is inversely proportional to the generated tension 'F', and also inversely proportional to the compression 'ξ'; then 'α' is (α = k / F = k / ξ). Consequently, the nodes generated by a highly compressed photon will be concentrated on a small part of the perimeter and vice versa.

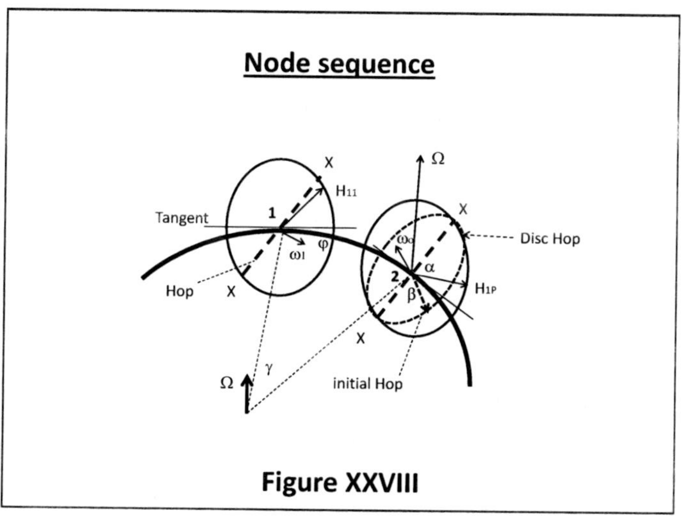


**Figure XXVIII**

The geometrical distribution of nodes on the ring perimeter depends on the photon compression but its directional distribution depends on the particle structural type. It is also important mention that the photon absorption is simpler when both particles have complementary Hyle structures that facilitate them fit appropriately.

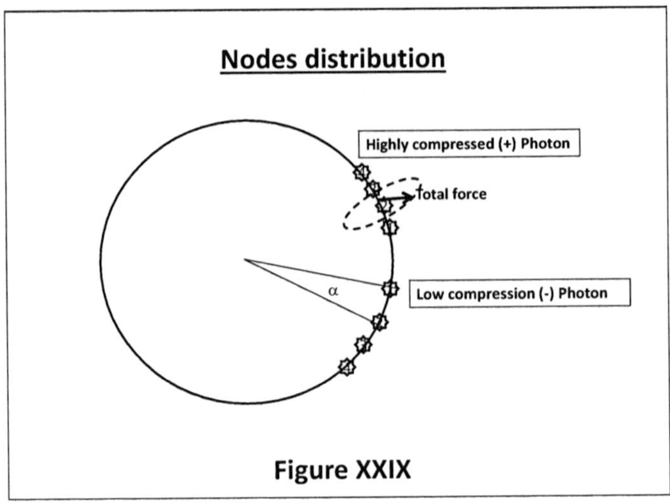



**Figure XXIX**

Another important aspect is that the force generated by each node depends on the direction of the missing 'H1' vector, so the total force will depend on the directional coincidence of the nodes; obviously when the nodes are closer the prevailing direction will be the same, so the total force in that direction will be larger when the nodes are closer.

The particle structure defines the type of effect that will generate the absorption; depending on the photon type, the ring will deform, will rotate or will break apart. Some of the main characteristics of nodes are:

- **The quantity** of nodes will be defined by the total asymmetry generated in the ring after absorbing the photon; this relation is given by the perimeter decrease which in accordance to Einstein-Planck relation is $[\Delta T_1 = h / \varepsilon_1 - h / (\varepsilon_1 + \varepsilon_o)]$. The unbalance will generate a total force that depends on the nodes distribution; it is independent of photon's energy.
- **Their distribution** is a consequence of photon asymmetry or compression. The effect of a concentrated distribution

is that all individual forces will point to a prevalent direction, so their addition will generate a larger force in that direction.

- **The magnitude of the total force** generated by all the nodes depend on several factors:
  - o The individual <u>direction of the force generated by each node</u> depends on the type of structure of the particle. Particle types V and R will have a tangential component while T type particles don't.
  - o The resulting magnitude of the total force for a given <u>direction</u> depends on the nodes concentration. This direction could be: a) tangential, producing the ring's rotation; b) radial causing ring's compression, or c) vertical producing ring's nutation.
- The directional <u>sequence of nodes</u> depends on the direction of rotation of the absorbed photon and the geometry of the ring; so, positively charged ring absorbing photons "Pp-" will roll towards the emitter and vice versa. The direction of the added force depends on the direction of '$\omega_o$' in relation to '$\omega_1$'.

## The origin of forces generated by nodes

Both the electrical attraction and the gravitational force exerted on massive bodies generate motion, but it is difficult assimilate how a very weak particle, as a photon, could produce powerful attraction and repulsion forces. Of course that the immediate answer is that the phenomenon is caused by a large quantity of photons absorbed by a large number of particles bound inside the body; but the correct reason is that the very weak photons generate large "real" forces by using the internal elasticity of the ring as the transformation media; <u>the internal energy of the absorbed photon resides in the AT</u>

<u>dimension, and it is converted in a force that moves the particle in the TT dimensions</u>.

Remember that force, by definition, is the action of elasticity of Hyle structures which recover their symmetry as ruled by the "Law of Symmetry"; so this is the basic mechanics used by nature; the internal elasticity of rings generates forces; obviously is a very efficient method for the generation of energy. This method converts the photon energy into mechanical energy which originates motion, or eventually is stored as internal mechanical energy (compressing the ring and tilting the discs) ready for deliver by the atomic structures when needed. Briefly; the "law of symmetry" origin of elastic force, generates <u>tangential forces</u> and <u>shear</u> whenever a node creates a gap in the particle ring by rearranging the angular sequence of its Hyle.

A detailed analysis on the internal mechanics of the effect of a node on the structure of Hyle is a complex issue that is out of the scope of this book. It deserves being studied in full detail, but a superficial analysis shows intuitively how each node cut the perimeter of the ring produces an instantaneous internal torque 'Tq' that increases the angular velocity '$\omega_1$' of '$H_1$' vectors, and simultaneously producing a centripetal force that stretches the perimeter. The nodes affect the direction of '$\omega_1$'[95]; and consequently, the tangential motion of Hyle on the ring's perimeter.

As mentioned before, the generated force may have three components: a) a radial centripetal component compressing the ring; b) a tangential component forcing the ring roll; and c) a perpendicular

---

[95] The ring with a node is comparable to a string of wool undergoing a twist in a spinning wheel; when just twisted it will tend to form a circle because the internal torque produces a radial centripetal force and also a tangential torque on the rope. The detailed study of this mechanics also deserves special attention for the consequences in the understanding atomic and molecular mechanics.

force (to the plane of the ring) applying a torque that produces a gyroscopic nutation.

The following analysis in this book assume that the type of structures generating the larger electric attraction, compared to gravity, use the tangential component of the forces generated by nodes; the addition of all tangential forces impulse the ring rolling in the appropriate direction; on the contrary, gravitational attraction is produced by concentrated radial forces which relocate the center of the ring.

This page is intentionally left in blank

## Chapter XIII

## Particle types: neutron, proton and electron

In previous chapters it has been shown that there are three possible configurations of particles, depending of the position of the rotation plane of its Hyle with respect to the plane of the ring. Two of them are charged particles, and the third is the neutron. There are two additional "antiparticles" with the same mass (size) of the proton, or electron, but with opposite charge. This chapter assigns a configuration to each one of them.

### Force orientation and particle types

In order to produce the tangential force needed to apply a torque that rotates the ring there must be applied a tangential force; consequently, the direction of the intersection line at the node "XXy" must have a tangential component; therefore, only the rings with a structure containing a tangential vector may be positive or negative particles; these structures are types V and R; not so the type T.

The following table summarizes the components of intersection line "XXy"

| Photon type - Particle type | Line XXy |
|:---:|:---:|
| Np/Pp – R | V + T |
| Np/Pp – V | R + T |
| Np/Pp – T | V + R |

The distribution of nodes is independent of the structure of the ring and is also independent from the direction of photon arrival; the intersection line "XX" depends only on the initial angle between '$H_1$' and '$H_o$' discs, although the initial angle with the tangent is defined by the direction of photon arrival and its internal structure, it remains constant along the periphery of the ring; this is why the nodes are placed on a specific sector of the ring.

Form the above analysis, we may conclude that the conventional sign of the charge assigned to electrons and protons will be the result of the emitter being a V or an R structure. In order to find which one is proton and which one electron, we must remember that protons share nodes with neutrons that are type T particles; both structures have virtually the same size, so contain approximately the same energy. In order to construct complex nets as demanded by the atom nucleus, they must share a node when both rings are approximately on the same plane[96]; this condition suggests that protons should be type V particles; consequently the electrons will be the remaining type R structure; this is shown in the next table.

| Particle | Structure |
| --- | --- |
| Proton | V |
| Electron | R |
| Neutron | T |
| Positron | V |
| Antiproton | R |

In order to facilitate the graphic analysis of the nodes amongst the particles, a visual and coded description of each particle type, follows.

**Proton: type V particle**

The following FIGURE shows a type V particle where the arrows represent static 'H' vectors on orthogonal positions around the symmetric ring. This structure outlines different geometrical

---

[96] A structure forcing to assemble rings perpendicularly, will not be able to build complex structures with tends of rings as commonly observed in the nucleus.

distributions of 'H' vectors depending on the reference phase (0 ° or π/2 in the FIGURE) at the superior end of the ring.

When the 'H' vectors in the disc rotate in opposite direction than the one that Hyle circulates in the ring perimeter, the structure corresponds to its reverse-particle. The FIGURE also shows how in this case the reverse-particle (or R-particle) never matches with the particle when applying successive rotations on its horizontal and vertical diameters.

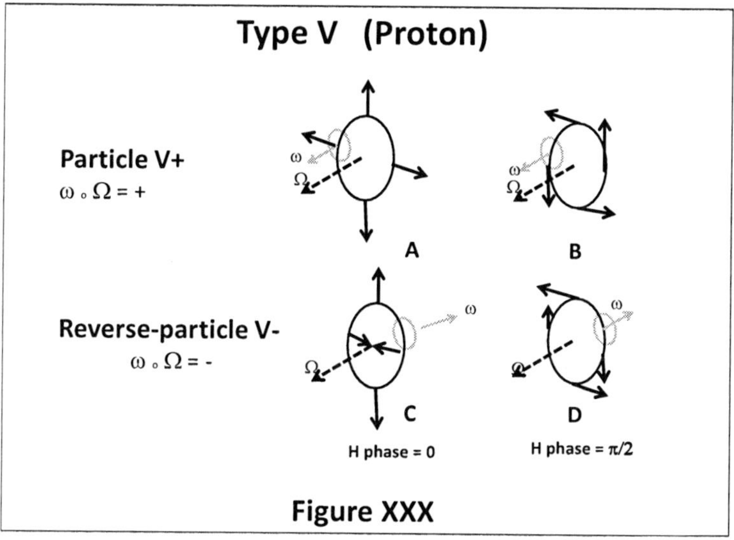


As explained in the following chapters type V structure is assigned to protons because their vectors 'H' may deviate slightly from the plane of the ring, so many nodes can be easily formed with several neutrons simultaneously, so knitting a spherical atomic nucleus.

## Neutron: type T particle

The T-type structure is assigned to a neutron because its structure lacks a tangential component of 'H'. The nodes in this type of particles generate centripetal and vertical tensions on the ring; so they do not force the ring rolling when absorbing a Pp.

This particle has its Hyle disc [97] perpendicular to the plane of the ring.

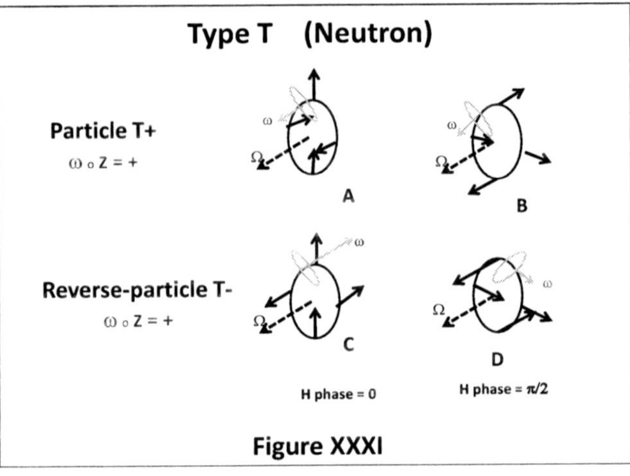



**Figure XXXI**

In next chapters will be shown how neutrons may support large asymmetries before breaking apart; so they are capable of absorbing large amounts of energy[98] before dividing into the protons and the electrons that populate the universe in similar numbers, forming atoms. This property makes them also difficult for detection because they do not emit photons until they break in two particles.

The asymmetry of the neutron is the key characteristic of this particle type, which enables it to produce stable charged particles.

---

[97] This feature suggests that these are the possible type of photons generated at the time of creation of the universe; but only one the T+ has been chosen. The circular shape of the neutrons ring, may be explained because the straight path has been deviated from the straight line because of some initially included shear.

[98] This behavior suggests that the neutron has a relatively stable non static structure; it can keep its ring shape while being asymmetrical, meaning that it does not emit a photon every time it absorbs one; simply increases its asymmetry accumulating energy until it breaks, producing two symmetrical particles, which are already static structures but in the case of electrons, they are very sensitive to any perturbing effect on its symmetry. The proton is also very stable; it does not break when absorbing photons, its Hyle structure will simply rotates faster when it absorbs additional photons.

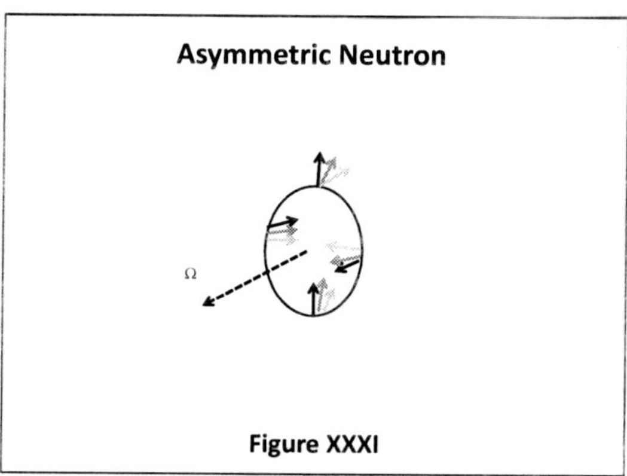



The asymmetry found in any neutron means that each 'H' vector on its perimeter changes its direction permanently, as illustrated in the following **Figure XXXI**, thus facilitating the formation of nodes with protons.

## Electron: type R particle

The next **Figure XXXII** shows a type R structure generated by a tangential Hyle disc; this structure is assigned to the electrons. This structure, as all others, may form two diametrically opposed pair of nodes with a similar concentric ring at any angle between them; this characteristic establishes "centrical" pairs. The FIGURE shows different shapes, depending on the initial phase (0 or $\pi / 2$) of the vector 'H' in the superior end of the ring.

The **Figure XXXII** assigns a negative polarity status when '$\omega$' which points to the center of the ring and also when pointing in opposite direction, because the particle and the R-particle are the same but viewed from a different perspective.

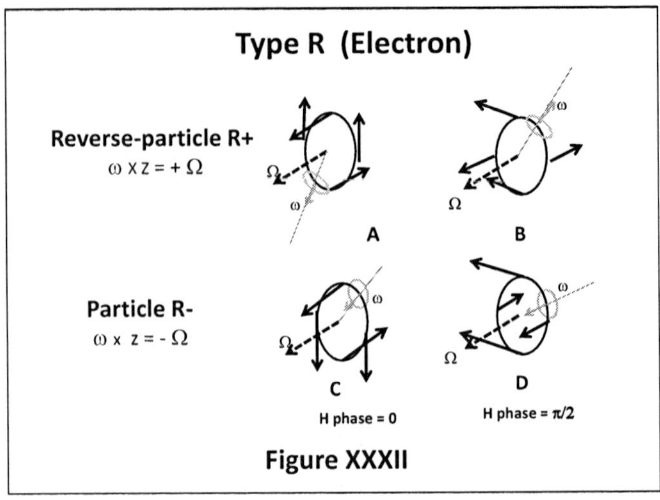



**Figure XXXII**

Type R particles emit Pp- photons with their vector 'ω' pointing to the opposite direction to the tangential motion of the Hyle in the ring; on the contrary, type V structures emit photons Pp+ with their angular velocity vector 'ω' pointing in the direction of motion.

## Nomenclature

It may be convenient use a nomenclature specifying the 'H' vector direction on each quadrant of the ring (1; 2; 3; 4), following a counter clockwise sequence, and using a Cartesian (x; y; z) frame of reference; indicating also the sign of the angular velocity 'Ω' on the "x" axis; as shown in the next FIGURE:

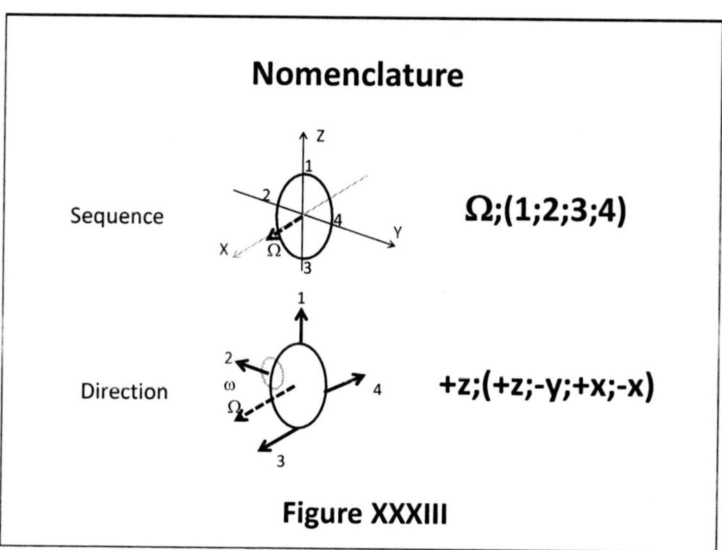



The following table describes the three types of particles and their antiparticles.

| Particle | Type | Ω | 0 | π/2 | π | 3π/2 |
|---|---|---|---|---|---|---|
| **Proton** | V | **+X** | +z;-y;-z;+y | -y;-z;+y;+z | -z;+y;+z;-y | +y;+z;-y;-z |
| R-Proton | V | **+X** | +z;+y;-z;-y | -y;+z;+y;-z | -z;-y;+z;+y | +y;-z;-y;+z |
| **Neutron** | T | **+X** | +z;-x;+z;+x | -x;+y;+x;+y | +z;+x;+z;-x | -x;-y;+x;-y |
| Neutron | T | **+X** | +z;+x;+z;-x | +x;+y;-x;+y | +z;+x;+z;-x | +x;+y;-x;+y |
| R- Electron | R | **+X** | +x;+z;-x;+z | -y;+x;-y;-x | -x;-z;+x;-z | +y;-x;+y;+x |
| **Electron** | R | **+X** | +x;-z;-x;-z | -y;-x;-y;+x | -x;+z;+x;+z | +y;+x;+y;-x |

## Nodes linking similar particles

Two particles of similar size may share a node when a point of both rings occupy the same position at the exact moment their 'H' vectors point to opposite direction at contact location; this node will keep them together until one of the rings changes its internal energy by absorbing a photon, impelling its Hyle rotate faster and so change diameter; by destroying the node.

Two particles of different size have different energy, so they will not share permanently a node because they rotate at different angular speeds (not always multiples); consequently, they will immediately go out of synchronism, applying a mutual stress that will end up separating them. For the TH, this is the basic mechanics of nuclear fission.

It is theoretically possible for two protons share a pair of diametrically opposed nodes, similar to an electron centrical, but it is highly improbable because there is no a natural force that may compel them share a common center, as the one given by a positive nucleus in the case of electron centricals.

For two particles of the same type it will not be easy to naturally approach slowly enough touch each other because they will repel when they cross the Sub-spaces E from the other; they can only get close if they approach from the right direction and at a slow speed.

The tangential node linking two similar particles, as the proton and neutron, is possible because the neutron is not attracted or repelled by the sub-space E from the proton, and the link is facilitated by the asymmetry of the neutron, whose Hyle rotates continuously when both particles touch at a low speed.

### Axial nodes among similar particles

Two rings of similar diameter may share nodes when one of them is slightly asymmetric, because a permanent phase change facilitates locking both structures at a given phase, and with two diametrically opposed nodes.

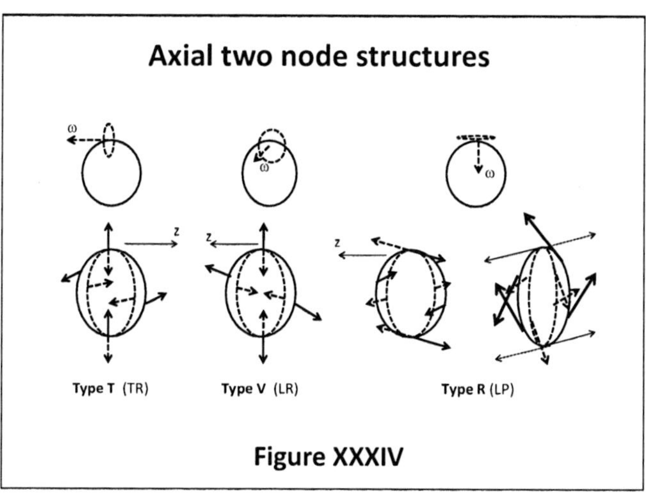



**Figure XXXIV**

The most important requisite for building a pair of axial nodes is for both particles must be concentric; this is easily accomplished when there is a centripetal force that maintains both rings in such position, as it happens in the atom where the very small protons attract the large rings of electrons placing them around the nucleus.

The two concentric structures may be pairs of the same type (V, T or R) because any of them may share nodes when an 'H' vector from one particle opposes the 'H' vector from the other, as shown in the **Figure XXXIV**.

The relative phase of the Hyle of a particle relative to the other determines the angle between the planes of the respective rings; the angle between centricals in atoms is defined by the correspondent proton ring plane in the nucleus.

## Tangential nodes connecting similar particles

Only two particles with similar diameter may share a tangential node because this characteristic defines the relative tangential speed of Hyle, and also its angular speed. The mechanics may be compared to gears that must be of the same diameter when all of them rotate synchronously.

The **Figure XXXV** shows a node linking a proton and a neutron; the two particles cannot form axial nodes because there in not possible place them in a concentric geometry.

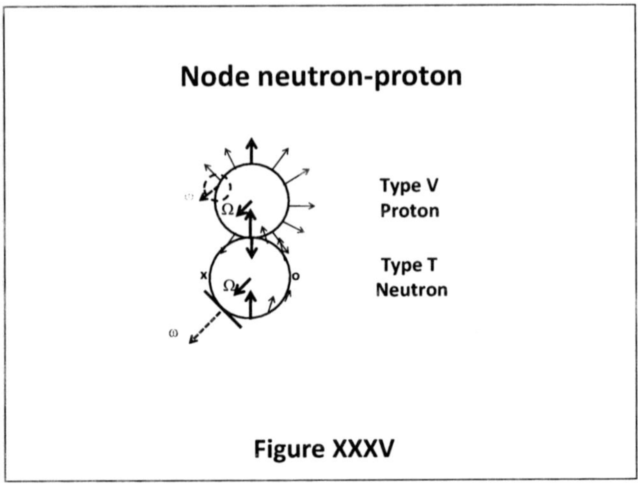


**Figure XXXV**

Although the **Figure XXXV** shows both rings on the same plane of the paper, they may share a node even having an angle less than $180°$ because the plane of the neutron discs may adapt to the most convenient direction because of its slightly asymmetrical structure.

## **Chapter XIV**

## **Detailed mechanics of absorption and emission**

Currently, physics theories offer mathematical descriptions resulting from the analysis of reasonable conjectures, which normally offer no more than a guess, and no serious explanation, about the ultimate cause of most basic natural phenomena, such as: how is gravity generated? Why matter bends space? Why the spin is a fraction? What is the origin of electric charge? Where is the missing energy of beta decay?

On the contrary, the TH proposal offers simple explanations to all of these problems; indubitably this theory, as presently exposed, is an incomplete attempt to explain everything, but definitely is a step forward towards such objective.

We, in the beginning of the XXI century, are fortunate inheriting the rich academic and experimental knowledge from the highly prolific observations and theoretical contributions made during the last century, so the labor in developing the TH was reduced to the labor of laying out some ideas in a coherent sequence; rescuing outstanding results from current theories, which made it easy propose a model that explains with remarkable simplicity the rich heritage of experimental data. The TH, as presented in this study, may be missing many explanations and mathematical development but it currently represents the only and most comprehensive theoretical proposal describing "nature's realities", ranging from "the real meaning of distance", the "description of atoms architecture", and "the universe formation at the beginning of times."

## Photon trajectory and discs orientation

It is important remember that Hyle is the substance constituting matter, so it has all the well known properties assigned to matter by classical mechanics laws as: elasticity, inertia, momentum, etc. The mechanical properties of matter are the consequence of one single law "The Law of Symmetry" which establishes that in all four dimensions, time should elapse at the same pace; any discrepancy manifests as the internal resistance of Hyle structures changing their length.

Initially, any just emitted photon will follow a tangential path; additionally, the tangential angle of its disc will be the same than that of the shearing component of the emitter Hyle at emission point; consequently, after completing the emission, the emitter will be freed from any stress, will adopt the most stable structure, and will  be symmetric again. When an extremely asymmetric particle releases its shear, may adopt a structural type other than its original one.

Other important fact is that the path of the emitted photon may be curved [99] when their discs have an angle '$\alpha$' other than $0°$ or $90°$ with the displacement direction 'z'. This is because the disc's vector '$H_o$' can be visualized as a chain of rotating rods, similar to a chain of gyroscopes[100]. The torque applied to the chain by the shear component of its Hyle induces the photon travel on a minimum stress path; therefore, <u>establishing the straight line geometry for the "space" it defines</u>. In the particular case of matter particles this path is a circle.

---

[99] The term "tangential" will refer always to <u>the instantaneous direction of Hyle displacement</u> at emission point.

[100] It is important to see that a pure geometric conversion is not applicable to a Hyle structures because '$\omega_o$' and '$H_o$' are not isolated vectors; '$\omega_o$' symbolizes the spatial direction of '$H_o$' at a <u>longitudinal position in the photon structure</u>, which means that the structure represents a cylindrical entity <u>with spatial dimensions</u>. Therefore, a rotation always demands applying a torque to cylinder axis. Let's imagine a type B+ photon as a rotating cylinder; if one tries to rotate it, will resist reacting with a precession movement; consequently, a particle does not represent to a pure geometric structure, it is a physical entity.

## Analysis of the emission process

Previous chapters describe how the photons have been divided in two categories: Type A which have '$\omega_o$' parallel to the displacement axis 'z' and Type B with '$\omega_o$' perpendicular to 'z'.

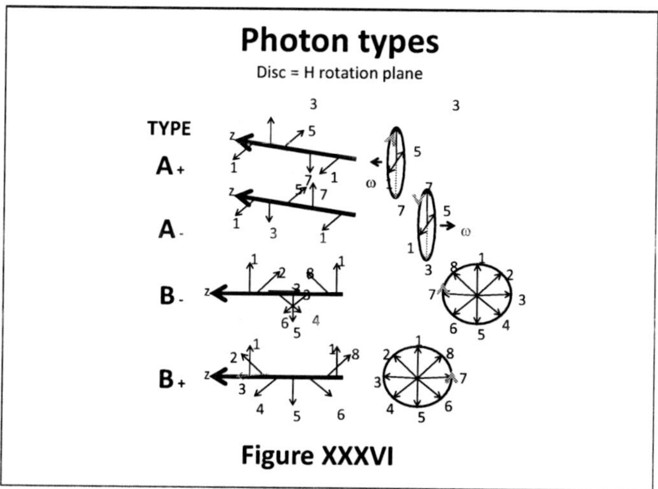


**Figure XXXVI**

As shown in **Figure XXXVI**, both types have two versions depending on the sign of the angular change when <u>the next sector of the structure passes through the same point in TT (space)</u>.

Every new node in a ring, increments the apparent rotation of Hyle because asymmetry gets higher as the sequential absorption process continues; the phenomenon is also influenced by the characteristics of the emitted photon, its period and direction of '$\omega_o$'.

The absorption process defines the geometric location and distribution of nodes, which were established by the coincidence of the pair of vectors '$H_o$' and '$H_1$' opposing each other on the intersection line XX. Therefore, it is reasonable to assume that the emission process follows a similar sequential protocol reconstructing the annulled sectors of the ring, by following a similar sequence.

In order to recover the initial symmetry, the procedure implies: firstly, re structuring the ring's geometry by "separating" the two opposite 'H' vectors, from the photon and the particle, which initially established the gap in the ring[101]. The emission process initiates when the shear generated by the nodes exceeds a limit, before the perimeter gap gets closed by the action of internal elastic forces.

The so described emission process that represents the sequential restitution of the missing 'H' vectors, may be imagined as the separation of the two vectors that produced that gap in the Hyle continuous function[102]; one of them releases the internal elastic tension by slowing '$\omega_1$' and simultaneously filling the gap in the ring's perimeter, while the other vector departs tangentially from the ring as part of a continuous structure; a photon. Therefore, the photon period and length are defined by the sequential release of '$H_o$' vectors from the rebuilt nodes when they reach the emission point at the ring perimeter.

---

[101] Rebuilding a node is equivalent to regenerate two opposite unit vectors which cancelled each other generating a spatial discontinuity in the Hyle function. To recover continuity, there are two options: a) leave to the internal elasticity forces to slowly close the gap, or b) to separate them by cancelling the weak energy bonding between the opposite vectors; this second case describes how a free type R ring (and possibly type V rings) will emit tangentially a Hyle function constituted by equal energy sectors represented by a chain of equally spaced rotating '$H_o$' vectors (discs), the distance between consecutive '$H_o$' vectors is defined by the next node to arrive to the ring's emission point after a number of revolutions; the nodes are sequentially rebuilt until the last cancelled node establishes the period '$T_o{}^*$' and the length '$T_o$' of the photon.

[102] The separation of two opposing physical entities from nothing appears to be an ontological trick; but in the case of rebuilding nodes one should consider that the energetic addition of two opposed vectors is a discontinuity in a continuous function that disappears when the function recovers continuity by liberating "the negative component" that generates the discontinuity. This logic could also be applied to the process of creation of the universe by generating a pair of photons departing from a point in opposite directions '$z$', and rotating at the same angular velocity '$\omega_o$'.

We could imagine any particle as a string of rotating rods or discs (following the useful segmentation model we have previously adopted to help us understand the absorption phenomenon); therefore, a node will be described as if one of the discs disappeared. Considering that the succession of discs shape a ring, regaining continuity involves tangential elastic forces which eventually may change the plane of rotation of the contiguous discs when their rotation axis is not perpendicular or parallel to the Hyle straight trajectory; when the discs axis point to other that those preferred directions, any longitudinal stress, applies a torque to its rotation axis; consequently the discs will react as gyroscopes acquiring a precession movement.

The proposed model could be applied to diverse particle types; but it is important to always keep in mind that the straight line in the own space of a particle, is a circle.

- In the case of a type R or type V particle, there is no such torque when the applied force is tangential, so the rotation axis does not forced change discs plane orientation; it will be like expanding or compressing vertically a chain of gyroscopes. When a small change of curvature induced by a node generates a non tangential force, may tear apart the Hyle structure; that is the phenomenon of of photon emission.

- Type T particles (neutrons) accumulate large amounts of energy from diverse photons coming from all directions, this process produces nodes distributed randomly, generating forces in all directions; this is why this particle may remain asymmetric during a long time before emitting an energetic photon. A neutron ring may <u>sustain diverse and dispersed tangential forces,  and radial/vertical shear</u> because they are distributed randomly on various perimeter points; so the cumulative action of forces in an specific part of the perimeter is a statistically random process; that is why free neutrons may have a diverse energy content.

- It was explained in the previous chapter how the sequence of nodes locate before or after the absorption point in symmetric particles, depending on the polarity (direction of rotation) of the absorbed photon and how its spacing depends on the photon compression.

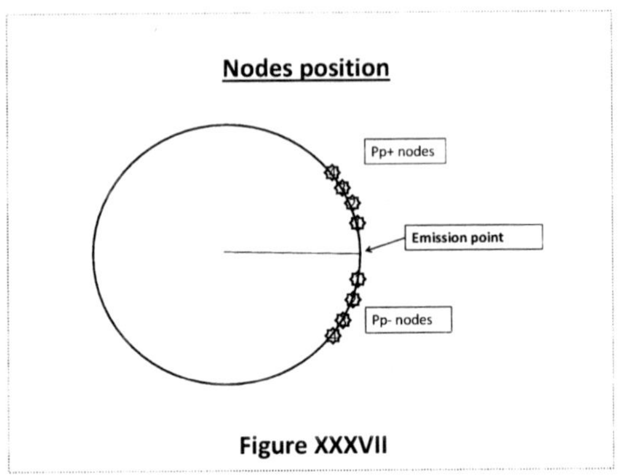



**Figure XXXVII**

It seems reasonable assume that the sequence of emission replicates the same order of node formation as illustrated in **Figure XXXVII**[103]; in most cases the nodes will be "behind" the emission point, so the angular speed of the emitted photon '$\omega_o$' will be set in accordance with the position of the group of nodes at emission point.

Assuming that protons are type V+ rings with '$\Omega$' parallel to '$\omega$' and pointing to the same direction; then will emit Pp+ (type B+ photons) as shown in **Figure XXXVIII**. Consequently, the electrons must be type R-

---

[103] Every new node emitting an 'H' vector will have to complete a number of revolutions before reaching the appropriate angle at emission point. This process explains why a smaller ring has to complete many more revolutions before completing the emission of a very long and weak photon.

particles with 'ω' pointing to the center of the ring; this particle type emits Pp+ (type B+ photons). This architecture explains how protons are positively charged particles which emit positive type photons (Pp+); and how negatively charged electrons are type R- rings which emit Pp- (type B- photons).

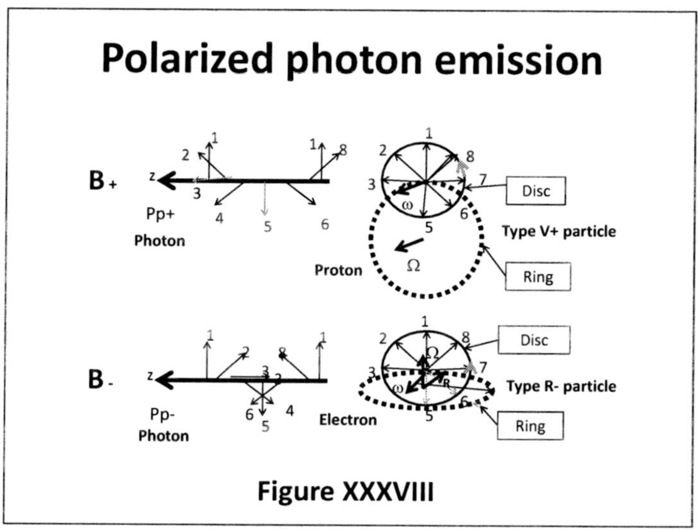


**Figure XXXVIII**

## **Internal forces generated by absorption**

In general, a particle is a symmetric ring of Hyle that <u>appears static</u> to any external observer because its temporal phase matches the spatial phase, so temporal and spatial oscillation are synchronized; but the particle stability demands an additional requisite, not being subject to any internal radial or vertical tension pushing the Hyle diverge from its circular trajectory. This kind of tension is generated when the gap generated by a node in any sector of the ring applies a torque for deviating '$\omega_1$' from its vertical "V" or radial "R" direction in the respective ring types, as shown in the next **Figure XXXIX**. When the emitted photon releases this torque, the disc plane recovers its original orientation, vertical or radial respectively.

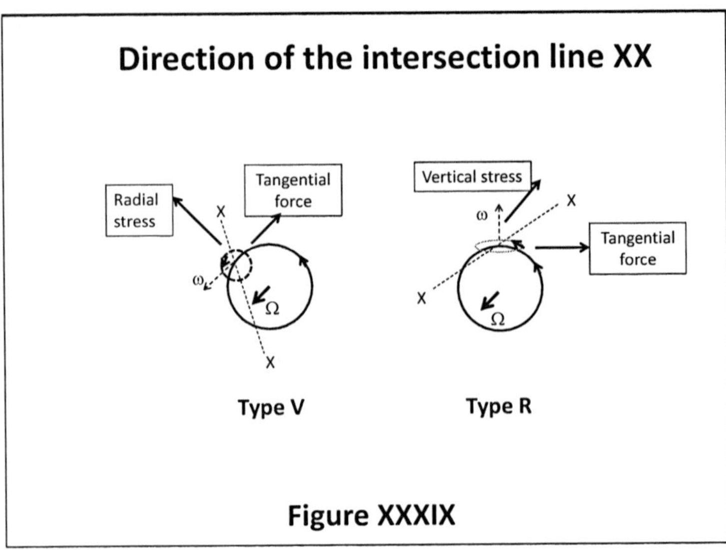



**Figure XXXIX**

When this kind of process is given, there are three consequences: a) The process frees the ring from the shearing tensions, so the particle discs return to their stable positions, b) The ring recovers symmetry by reducing its internal energy; and c) The angle '$\alpha$' of '$\omega_0$' and the tangent at emission point is the angle of the straight external trajectory of the emitted photon, so the shear tension deviates the photon trajectory until its discs recover an orientation with no shear[104]. The photon relieves the ring compression by taking away energy and inherits the shear because every '$\omega_0$' vector leaving the particle will be oriented in such a way that there is no shear left in the particle ring; this process suggests that all photons inherit a shearing force, which curves the initial portion of its tangential trajectory, until the discs adopt a stable Pp or Np structure, as it will be shown in the analysis of beta decay.

---

[104] The angular velocity '$\omega_0$' from the photon discs leaving the ring will not have a null angle with the tangent, but it will adopt the most stable structure as they increase the TT. In this way emission process generates Pp or Np.

The direction of the intersection line XX remains constant during the absorption process. This direction is always initially set by the discs plane in the particle and the angle of photon discs.

## Energy storage

In previous chapters we have seen how a photon decreases the period '$T_1{}^*$' of a particle that absorbs it, while its elasticity slowly decreases the perimeter '$T_1$' of its ring, under this circumstances, recovering symmetry by the action of internal elasticity may last for a long time. The only way preventing the rupture of particles subject to this kind of stress is by applying a strong external compression force predominantly tangential which closes the gap faster, preventing it breaking while asymmetry increases. In this case, the ring decreases its diameter faster than it increases its energy, until <u>reaching a new symmetric status by increasing its internal energy content</u>; the pre-compression prevents expelling surplus energy as a photon because the perimeter remains smaller while additional energy is absorbed. This simple analysis demonstrates that is possible have free <u>protons or free electrons as stable static rings but with larger internal energy than the nominal one</u>, when they were pre-compressed while absorbing a photon. This basic mechanics allows static particles share metrix in any stable environment; such as atoms having stable electrons with diverse energy levels around a nucleus, or nucleus increasing their energy in accordance to the surrounding metrix. Even free electrons or free protons not necessarily have to have the same nominal (or minimum) internal energy as currently assumed[105]; they may store some energy because they were generated while they were subject to a considerable compression; the excess of energy is released when it breaks, reaching then the minimum energy level for a free particle of that type, in a flat space.

---

[105] It will be shown that this phenomenon may explain the energy differences found in Beta decay process, currently attributed to neutrinos.

An important conclusion is that in both types of non T type (V and R) particles recover symmetry after absorption is because tangential forces applied to anyone of them will produce a perimeter change which only affect phase/period changes in Hyle rotary structure; phase shifts may be compensated by applying a tangential compression which preserves its "straight" trajectory.

The total tangential force applied to a ring is a function of the quantity of nodes and their distribution, but it will always be proportional to the compression of the absorbed photon; therefore will also be proportional to the emitter mass. A highly concentrated node distribution increases the tangential component because the spatial phase difference among them is minimized; this characteristic explains how a nucleus keeps around it, the stable electrons with higher energy content.

## The metrix of particles

The metrix of any symmetric free particle in a flat space is equal to 1. This symmetric condition is also achieved by particles structuring a massive material body which emits highly asymmetric photons; this is achieved when particles are subject to the same mechanics of pre-compression while absorbing energy, as described in previous paragraph. Consequently <u>the curvature of a symmetric ring</u> (nominal mass plus stored energy) is the parameter defining the metrix of the photons it emits, and consequently the metrix of the sub-space its emitted photons generate.

Therefore, the metrix of a symmetric emitter does not refer to the ratio of the particle perimeter '$T_1$' and its period '$T_1{}^*$' (as defined for photons and sub-spaces), but it relates with the extra curvature the ring acquires by exhibiting an extra compression it with respect to the nominal period '$T_n{}^*$' ($\mu = T_1 / T_n{}^* = T_1{}^* / T_n{}^*$) of that particle; it can be symmetric but smaller.

A particle may still be a static structure while resting in a fixed position on a sub-space [106] whose metrix is defined by compressed photons crossing that location, even when it is subject to the gravitational and electrical forces communicated by these signals. In order to achieve a symmetric stable condition the ring should be subject simultaneously to a strong external compression force while it increases its internal energy; the final result will be that the particle contains some extra internal energy and a reduced diameter; consequently, it will emit compressed photons with the metrix set by its new mass, energy, and the curvature of its ring.

The increase of its internal energy is conventionally known in classical mechanics as "potential energy." This process may be illustrated by visualizing a ring resting on a rigid fixed surface which maintains its position in reference to the emitter, so it resists to the action of the internal elastic force generated by its nodes, moving it closer to the emitter. The generated compression may be applied to the ring directly by a node, when the absorbed photon generates a tangential force that compresses the ring. This is the case of centrical electrons which are compressed by the highly asymmetric photons [107] emitted by the very close nuclear mass.

A symmetric ring is not a stable and static structure[108]; it increases its asymmetry with every new photon absorbed, because the elastic

---

[106] When a particle is symmetric it is always a static structure, and when additionally does not move in the sub-space it is said that the particle resides in that location of the sub-space, sharing the spatial metrix at that point; it does not move because there must be an external force impeding it.

[107] As explained before, the compression force applied to the centrical ring is always tangential; additionally all the nodes agglomerate close to absorption point because of the compression of arriving photons. The '$\omega_1$' vectors in the ring, do not change their tangential orientation in reference to the circular straight line defined by the own space of the particle.

[108] Strictly speaking, an asymmetric ring cannot be considered as solid matter anymore, because the intuitive definition of matter implies a permanent shape, mass and volume; which it loses

tangential component of the force generated by the node does not close the gap immediately, but the magnitude of this internal compression could be enough in order to restrict the breakage of the ring; so the particle will keep storing additional energy. This characteristic explains how a cluster of diverse objects share a common metrix (compression).

A detailed description of the mechanics of asymmetric photon absorption that changes the metrix of a particle follows:

1) A particle is always an elastic ring closes any gap in the perimeter by applying an internal elastic force that closes this discontinuity generated by a node when absorbing a photon.

2) A compressed Pp ($\mu < 1$) distributes the nodes along the perimeter in such a way that the ring, rolls towards the emitter; a compressed Np moves the center of the ring, do not rolls it.

3) When an external influence prevents the particle move, the diameter of the particle will decrease its diameter a little with every new photon arrival, until it gets the smaller stable diameter for achieving symmetry ($T = T^*$); in other words, the external influence grants enough time so that the slow elastic process, closes the perimeter gap.

4) Contiguous energy packets in a compressed photon arrive at time intervals equal to ($\Delta t = T_o^* / n_o$; and: $n_o = T_o^* / T_1^*$). The particle, after absorbing enough asymmetric photons, gets a new period ($T_2^* < T_1^*$) so the photon energy packets, arrive to the same point in the ring perimeter; in this way, the particle becomes symmetric

---

when becoming asymmetric; therefore, an asymmetric ring may be more properly described as a photon travelling in a circular space.

after decreasing its perimeter; the internal elastic force also decreases as the ring reaches symmetry; obviously the external reaction which impedes it moving will also get smaller.

5) When the particle achieves symmetry positioned at a fixed location, it will have the same metrix of that point in the sub-space it resides; therefore, any new photon arriving to it increases the energy but does not decrease the diameter of the ring anymore, disturbing its symmetry, so the particle releases some energy, in the form of a photon.

6) Whatever the energy of the emitted photon, its compression will be proportional to the actual curvature of the ring; originated by the geometric change from a Circular space to a Cartesian one.

7) When the external force is zero, the particle will accelerate towards the emitter, but will keep its period and its perimeter ($T_2 = T_2^*$); if eventually it is forced to stop in a point closer to the emitter, it will get smaller and adopt the new metrix; on the contrary, if an external force takes it far from the emitter, the particle will recover its nominal period '$T_1^*$'.

In the following chapters will be explained how the internal mechanics of nuclear protons enables them share the cumulative metrix, and why the photons emitted by them are also compressed proportionally to the total nuclear mass.

## Mass-energy equivalence; wave-matter duality

It is very important analyze the real meaning, and restrictions, of the classical concepts of inertial mass; gravitational mass and energy, under the scrutiny of the TH:

- The gravitational mass from an emitter is defined by the effects of the initial compression of the emitted photons established by the diameter of the emitter ring '$T_1$', which is the same than its period '$T_1{}^*$' for symmetric particles structuring a material body.
- The metrix of the sub-space G generated by an individual particle is established by the compression of emitted photons; which as previously shown, initially is proportional to the emitter mass.
- The inertial mass is defined by the amount of absorbed energy by a particle in order to increase its speed in a velocity field.

Emitting low energy photons is not a universal characteristic of particles; for instance, a free neutron can store large amounts of internal energy but cannot generate a sub-space because it normally does not emit weak signals. Moreover, when a neutron resides in a sub-space, it is not always moved by the forces generated by it, although increases its internal energy, and what is more important, its asymmetry with every absorbed photon, but it creates nodes randomly dispersed. A neutron is an extremely asymmetric ring, so any new node generated by the absorption will be randomly distributed along its perimeter, so all ring sectors will be homogeneously compressed; consequently, the total force generated will not have a predominant direction which may force the ring change its position.

Consequently, considering that a neutron is not a natural emitter of low energy photons cannot generate a sub-space G (gravitational field), although it has an internal energy that in accordance to classical paradigms should be equivalent to a mass. The classical mass concept of an material object is not applicable to a particle under this situation; however, the particle contains energy as defined by

Einstein-Planck relation, but as it is asymmetric it is not a material object with mass, neither a photon that travels out of an emitter.

Under conventional human semantics, one may say that the fundamental property of matter is the particle ring being a static stable object that remains being the same indefinitely and, independently from the surrounding environment, until it is perturbed by the arriving signal which always destabilizes the ring converting it from matter to wave, because under this perspective, an asymmetric particle is a dynamic four dimensional entity more appropriately associated to a wave moving in a circular space than the static entity that defines a material particle. Curiously, conventional physics theories offer a conjecture about this reality instituting the "principle of matter-wave duality" [109] as a difficult understand but confirmed "natural fact", arriving to this conclusion after <u>interpreting it subjectively as a firm evidence of the suposition</u>. This is not the only case in science; there are many observed phenomena, whose interpretation is strongly influenced by the vision imposed by paradigmatic principles.

A symmetric ring is "stable" from the point of view that it appears difficult change, but this is the result of its brittle nature, because it is extremely sensitive to shear; the ring emits any small energy surplus when its symmetry is slightly perturbed, as observed in the case of free protons and electrons.

Moreover, strictly speaking, the diameter (mass) of any free ring is not a universal property for all particles of that type; there may be free electrons containing additional energy because they were initially

---

[109] There is a price already paid by science for the adherence to the matter-wave duality as an "act of faith" (since it does not offer an explanation other than the interpretation of an observed phenomenon). It is normal to find scientists who do not trust human logic or common sense, and accept as real the most bizarre results, obtained by mathematical manipulations applied to incomplete causal relations.

subject to strong compression force which prevented them breaking apart while they were absorbing some extra amount of energy[110].

Consequently, there may be particles (symmetric rings) of the same type with smaller diameters (larger mass, more energy content) than the typical particle. This particles will emit energetic photons with a slightly different compression because of their mass difference, as in the case of released electrons occupying an centrical with a considerable amount of energy; they emit photons with compressions defined by the radius of the centrical (classic concept of gravitational mass), but the energy of these photons will be_proportional the difference between the initial and final particle period 'T$_1$*', delivered when the particle changes centrical energy levels, or when they become a free particle with a period 'T$_n$*'.

Most free particles are not pre-compressed while absorbing photons; therefore, their observed energy content is close to the nominal value; therefore, it is useful, but imprecise, assume that every particle has a predefined mass.

A free receiver in a sub-space changes its location because the total internal elastic force generated by the group of nodes has a prevalent direction. Grouping nodes in the perimeter is proportional to the ring

---

[110] It is theoretically possible to artificially produce new negative particles much more energetic than an electron by applying high compression to its ring while it absorbs energy; but one must count with a mechanism equivalent to a massive nucleus and some way to liberate the energetic electron without disturbing it afterwards. It is important to note that the energy of a photon is hundreds of thousands times smaller than the energy of any stable particle, so the asymmetry induced by a node is very small. Considering that the absorption time is given by the length 'T$_o$' of the photon, the absorption process lasts while the receiver ring rotates hundreds of thousands of times, so when the tangential force is strong enough to close the node gap, it will prevent ring rupture; this is the case of an electron in an atom. Moreover, a free particle may emit a tangential Pp while simultaneously absorbing an external photon generating strong shear on the particle perimeter; this may be the case of a free electron moving towards an emitter while sending a Pp- of equivalent energy to some other direction, generating its own individual sub-space E.

curvature (or its mass "m"), and to the absorbed photon compression, which is proportional to the emitter mass "M"; these characteristics explain clearly the reason why in Newton gravitational law ($F = G M m / r^2$) the force is proportional to the emitter mass, and also to the receiver mass.

In the next chapters it will be analyzed how a nuclear neutron acts just a useful entity gluing protons, and also adds mass to the nucleus because it recovers its symmetry, therefore, its condition of material entity.

**This page is intentionally left in blank**

## Chapter XV

## The neutron

### Neutron stability

The neutron is the mother of all particles in nature. Possibly, neutrons were generated at the beginning of times by a powerful asymmetric continuous gamma ray emission which evolved forming rings [111] by wrapping on a circular path, caused by a possible random orientation of their discs; these primordial neutrons broke generating all kinds of secondary particles by collision and absorption. Therefore, neutrons are very stable asymmetric type T rings able to absorb a considerable amount of energy from arriving photons, until they break out generating the emission of a high energy asymmetric photon with a tangential '$\omega_o$' angle '$\alpha$' ($0° < \alpha < 90°$).

The TH suggests that actually very few neutrons emit low energy photons, because their rings normally resist a large shear. This behavior makes them behave as energy reservoirs which break apart only when the internal tension generated by the asymmetry and the resulting shear exceeds a limit[112]. After the emission process finishes, a static neutron or other symmetric charged particle (a proton or an antiproton) is left behind; in most cases, the powerful asymmetric photon emitted, generates a stable and symmetric electron (or

---

[111] My guess is that is quite improbable for antineutrons to exist, because the primary high energy photons must have been constituted with Hyle rotating in only one sense, travelling in opposite directions.

[112] Generally, neutrons should not be able to generate a sub-space because normally they do not emit low energy signals. It is reasonable to predict that even in the case that a star collapses forming a mass containing only neutrons, it will be intrinsically unable to generate stable gravity because it will not emit enough low energy signals capable of communicating its existence to the surrounding environment.

positron) containing more internal energy than the nominal energy of a free electron in flat space. Quite possibly, this is how nature creates a "neutral" universe generating most of the time an equivalent quantity of protons and electrons, which automatically get together constituting atoms.

It has been shown before that type V particles, and especially type R particles, are very fragile; they cannot stand shear, so they are prone to emit a weak photon recovers from small asymmetries; any small non tangential force component may force Hyle diverge from its circular path, breaking the ring and liberating any energy surplus as a photon. The stability of neutrons is a consequence of the tangential angle '$\alpha$' of '$\omega_o$'; moreover, type T structures may store considerable energy before fracturing because they disperse the direction of forces generated by nodes, as their ring become increasingly asymmetric. When the internal stress reaches a limit its surplus energy is so large that in order to recover symmetry the ring must release an extremely energetic photon[113]. The geometric details of this process are analyzed in the next subtitle.

Recapping; neutrons cannot generate a sub-space G because they do not emit low energy photons. This is why a neutron is not a natural emitter able to communicate its existence to the universe. Free neutrons are energy reservoirs with a contingent mass[114] unable to generate a gravity field. This particle when free may move undetected within a sub-space; it is only detectable when being in an atomic nucleus; one may learn about its existence indirectly by detecting the particles it generates.

---

[113] We will define "nominal energy" of a free stable particle as the minimum energy level that it has when resides in a flat space. For some unknown reason there are not stable particles weaker than electrons, neither a weaker electron. Probably weaker particles do not subsist since even a low energy photon may break them.

[114] It was explained previously why an asymmetric ring is really a photon travelling in a circular path, not a material static particle.

As a free neutron does not navigate on the velocity field (gravitational or electric) because it is an asymmetric structure, it does not behave as a typical matter particle. It is really a circular photon.

The direction of the intersection line XX between a neutron disc and the disc of an external photon generally will be <u>perpendicular to the tangent of the ring</u>, so it will generate a shear with every new node[115].

Most neutrons in nature are not symmetric rings, so they are not strictly neutral particles because they have always some extra energy stored, meaning that their discs are always slightly tilted in some of the three directions (T, V and R) that characterize the "charge" or "no charge" quality of particles; consequently, all neutrons are a blend of a neutral, positive and negative particle because of the induced asymmetry generated by photons arriving from diverse directions, tilting the angular velocity '$\omega_1$' randomly, so changing its relative magnitude in all three axis (T, V and R).

This characteristic may offer a possible explanation on how a nominally neutral particle as the neutron, may respond to the directional influence of magnetic fields[116].

---

[115] These arguments explain the conventionally accepted natural instability of neutrons by detecting their short lifetime.

[116] One must remember that a magnetic field is generated by geometric devices expressly assembled to apply a directional electric force at a particular point of a sub-space. Obviously, the influence of this stable directional force on a type T ring will tilt its discs, increasing some of the R or V components; consequently the particle will react as a small charge while subject to this kind of external influence. It will be interesting to investigate if the three components of '$\omega_1$' in neutrons may be related with the "quarks" proposed by Quantum Theory.

## **Neutron rupture; beta decay**

In order to break a neutron apart, a large internal shear force must be applied to the ring perimeter after it stored a considerable amount of energy. When a type T structure breaks, releases a high energy photon which closes on itself so constituting a type R ring, eventually containing more energy than the average free electron. The distance at which photon closes is a function of the relative magnitude/direction of the initial shear component at emission point, and the amount of surplus energy at that moment.

For a not so obvious reason the neutron normally changes its internal structure from a type T to a type V particle after emission, converting it in a positive charged particle, a static proton free from any shear. This important phenomenon is the key opening the door for the creation of a complex universe, by splitting apart this mother particle, so an equivalent number of electrons and protons are generated, which afterwards; automatically constitute atoms; assemble molecules, and more complex material entities.

The distance from the emitter to the point where the photon closes on itself forming a ring, depends on how far the emitted high energy photon travelled at light speed, before structuring an electron. This travel time possibly responds to a non linear process, which obviously is a function of the relative magnitude of the inherited shear component, and other factors as the direction of emission, and the emitted energy. Any passive detection method, capturing the emitted electrons at a given distance should account for all these factors[117].

---

[117] The conjecture proposed by Pauli and Fermi, suggests the existence of neutrinos as the destiny of the extra energy; while under the TH perspective it is contained inside the electron. It is important to note that the existence of neutrinos generates a paradox; a small particle means an entity containing a large amount of energy; and on the contrary, a low energy particle will always refer to a lengthy entity; so neutrinos cannot be "small and low energy particles" at the same time.

The observer should be aware about the fact that the energy of these electrons may deviate from standard values, and must also be very careful in applying the concept of "kinetic energy" to this complex transformation process involving a photon that travels away increasing the TT at light speed before evolving to a smaller electron (higher energy), moving at non relativistic speeds. **Figure XL** illustrates this phenomenon.

To understand the beta emission process under the TH perspective, one should be aware that the photon curves its trajectory as a consequence of the magnitude of the tangential angle '$\alpha$' of '$\omega$'. The process of curving the photon path [118] cancels '$\alpha$' by setting the circle as the straight line in the own space of the particle.

It is also important note that the diameter of the electron is two thousand times larger than the proton diameter, so the photon travels away at light speed for a considerable time before closing as a ring; so preventing that the just created electron ring, surrounds the emitter. Therefore one may conclude that the generated electrons emerge at long distances from the emission point, indicating a slow and complicated wrapping process takes place at distances equivalent to many electron diameters.

---

[118] Obviously this description is approximate because the natural phenomenon involves distortion of a continuous media; it would be more appropriate describe it as the process of applying a torque to a metal plate to curve it.

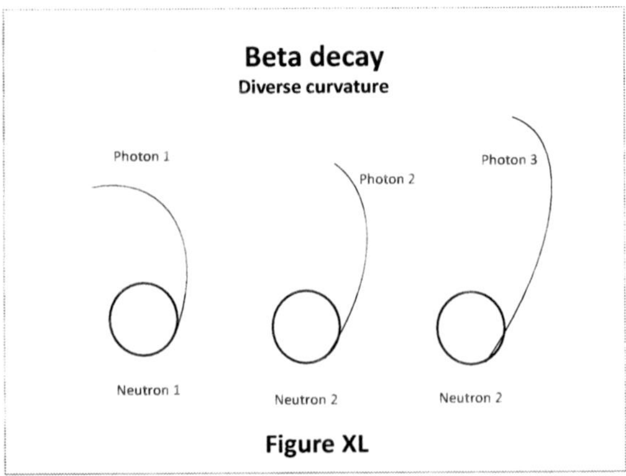



**Figure XL**

As explained before, the absorption of diverse energy photons arriving to a neutron from all directions will generate an increasing asymmetry on the ring, but the force generated by each node will be oriented in diverse directions. Considering that somehow the vertical component of the angular velocity of the discs ($\omega$) remains in the emitter ring, then the angular velocity '$\omega_o$' of every disc from the just emitted photon has only tangential (T) and radial (R) components (relative to the plane of the ring). Once in the photon, the tangential disc (T) will become radial as the photon trajectory is curved by the radial (R) component the tangential disc (T) will become radial as the photon trajectory is curved by the radial (R) component.

The following discussion explains some basic ideas on how the emitted photon may coil circularly, generating electrons [119] (type R rings), and how the Hyle in the neutron structure splits in two entities:

a) The highly asymmetric neutron composed by substantially tilted discs, evolves to a type V ring when it gets free from the shearing components[120].

---

[119] It is quite reasonable to assume that under special circumstances the tangential angle '$\alpha$' of '$\omega$' will have a relative magnitude that the neutron will expulse a positron, leaving an antiproton. The emission process details should be quite similar than in the one described above, but quite probably the just born antiproton, and the released positron will be destroyed immediately by nearby protons and electrons, producing powerful gamma rays.

b) The asymmetric high energy photon leaves the ring with an angular velocity '$\omega_o$' and a predominant tangential (T) component parallel to 'z', and a radial (R) component pointing to the center of the emitter ring at the emission point; these components are released when the ring fractures.

The emitted photon that describes a curved trajectory driven by the radial component (R), closing circularly, is a type R ring (electron), as shown in the next **Figure XLI** "Neutron Decay."

The structural change in the emitter ring may be explained by decomposing the angular velocity vector '$\omega 1$' of an asymmetric neutron in its three components (V, T, R); so this particle could be described as containing three overlapping rings. When this multiple ring assembly breaks apart, it will keep the most stable structure (V); sending away a photon that carries out the shearing components (T and R). The T component is more important than the R component, because initially the symmetric neutron was a pure T structure before evolving to a highly asymmetric hybrid ring; consequently the most stable component will be T; in other words, the photon will have its angular velocity vector '$\omega_o$' forming a small angle '$\alpha$' with the displacement direction 'z', because this angle will be proportional to the relative magnitude of the R component, which will be a function of the surplus energy accumulated in the neutron before emission. The physical effect of this angle on a photon structure is that the trajectory bends, until closing as a circle <u>forming a Type R- ring, with '$\omega_o$' pointing to the center of the new particle.</u>

---

[120] Although, sometimes could evolve to a type R ring; leaving an anti-proton.

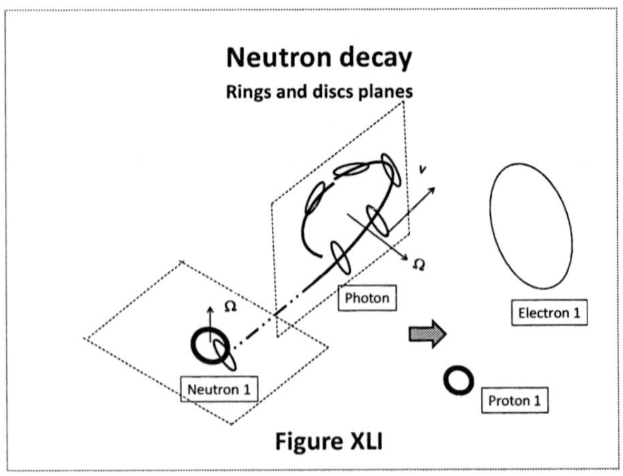



**Figure XLI**

Another visual description of the phenomenon may consider the emitted photon as a chain of individual gyroscopes moving on a straight path (in the circle), each one represented by a rotating disc, travelling on the tangent of the ring; as shown in FIGURE XLI. The angular velocity '$\omega_o$' at emission point has an angle ($\alpha < \pi/2$) with 'z', suggesting that the radial component '$\omega_r$' will apply a torque to the gyroscope rotation axis, so producing a vertical precession which bends the trajectory vertically. The curved path[121] defines the straight line in this individual sub-space; finally the photon structure will close as a circle, resulting in a type R- ring with a circular own-space geometry.

This figure explains the most probable behavior, when the tangential component '$\omega_t$' of the angular velocity is predominant, and the radial component '$\omega_r$' is relatively small. When '$\omega_r$' is larger than '$\omega_t$', the photon may evolve as a positron.

---

[121] A gyroscope with a horizontal rotating disc which is moved vertically or horizontally will not generate any precession; only applying a torque which changes the direction of its rotation axis may do it. It is important to mention that as the own-space defines the circle as inertial path, there will not be any torque applied to the rotation axis in a type R ring.

It is important note that the new ring plane is perpendicular to the plane of the emitter; to wisely preventing that the just born electron touches the sub-space from the proton, and may get trapped.

This simplified proposal needs being experimentally and theoretically redefined. This phenomenon could be mathematically analyzed in three different ways:

- Describing the neutron as a circular chain of discs; so the initial type T particle increases its asymmetry as it absorbs more photons, tilting the discs, which generates the internal shear which ends up breaking the ring.
- Considering the neutron as a continuous circular laminated material which releases stress by peeling the shear components. It is important mention that the length of the emitted photon is approximately two thousand times larger than the ring perimeter, so the ring must complete more than 2000 laps during emission; under this circumstances, the photon may be peeling more than 2000 extremely thin layers; every released layer increases the diameter of the ring by decreasing its radial compression, and releasing the shear.
- By performing a gross, non geometric analysis, similar to the classical approach: measure the initial energy of the neutron and the remaining energy left in the two final particles; the proton and the electron. When analyzing the beta decay of nuclear neutrons, one must take into account the additional energy of neutrons tied to protons in that particular atom.

Other type of unstable particles, as Muons, may result from photons evolving into rings containing large shear components which shatter them apart in a short time.

**<u>This page is intentionally left in blank</u>**

## Chapter XVI

## Atomic Structure

This chapter explores and explains how the few fundamental laws proposed by the TH may describe the simplicity and harmony of the basic atomic mechanics. After consolidating the TH predictions herein presented, the next stage should explore mathematically the phenomenon, adapting the classical equations to the processes described in this chapter; task well beyond my personal potential, which additionally falls out of the scope of this study; however it offers an interesting task to future research.

This chapter describes how the atom is assembled; how the nucleus is structured as a knitted basket of protons interlaced with neutrons, where every proton has a particular flat sub-space E that traps electron rings and places them around the nucleus by storing energy in them. The electron in the atom will be known as "centrical", substituting the classical concept of "orbit", and marking a difference with the name of "orbital" given by quantum mechanics [122] when describing the energy levels from atomic electrons.

When a given proton in the nucleus does not have an electron occupying its sub-space; a Pp+ emitted by a proton travels away from the atom setting a sub-space E+, the Pp+ is not absorbed because there is not a surrounding ring; so it may intercept and attract any free

---

[122] The orbit in classical mechanics is the path followed by a rotating object. The orbital in Quantum Theory is the energy level containing an electronic "cloud" which results from manipulating mathematically the electron energy distribution. The TH offers a clear explanation unifying coherently both concepts. We use the term "centrical" by defining it as the electron ring that surrounds the nucleus residing on a particular proton sub-space E+, whose diameter defines its particular energy level.

electron within its reach; rotate the plane of its ring, and compress it in order to store the largest amount of energy possible. It also describes how a centrical electron, emits an Np when changes energy levels. It is interesting note how the highly subjective conjectures proposed by De Broglie, and the Schrödinger equations, fit so closely to the descriptions proposed by the TH model. The conventional approach focuses on suggesting mathematic causal relations but not being able to explain the basis of the observed phenomenon. The protocol followed by the TH is absolutely different; it describes the phenomenon as the logical consequence of its fundamental proposals, and thereafter explains its causes.

Initially, this chapter presents a brief analysis on the emission process and its relation with the particle type; it also apply these concepts to the investigation of neutron rupture explaining alternative explanations to the hypothetical diversity of kinetic energy observed in the emitted electrons. It also investigates some other phenomena as: the neutral and polarized photon emission mechanics; it suggests some explanation about the existence of the "magic numbers"; it also analyzes how a positive sub-space E is produced when some electrons miss in the atom; and how nature integrates the nuclear particle masses in a material object which generates a gravitational field proportional to the consolidated mass.

In summary, the following paragraphs explain how nature automatically creates complex atom structures by just observing the few basic TH laws. At the beginning of times, light evolved to neutrons which divided in equal number of protons and electrons, automatically composing atoms, molecules and all the material objects that fill the universe on which we reside.

I devoted my time mainly, investigating the ideas contained in this book; I avoid getting involved in digging more deeply on the details and consequences of the area investigated; but I did not observe the rule in this chapter; just showing the evidence of the absolute coherence of the TH proposals with nature's reality.

Before entering to the analysis of the atomic structure, there are some concepts about the geometric characteristics of photons and rings that must be investigated in more detail.

## **Proton**

This particle is a symmetric type V ring which emits Pp+ photons when the absorbed energy exceeds a limit. The sub-space E that it generates is always on the same plane of the particle ring.

As it was concluded before, proton is the side product left after an asymmetric neutron breaks apart emitting a powerful photon.

The protons have similar energy and size than neutrons (0.55 % nominal difference), this is why they can share stable nodes when their 'H' vectors point to opposite directions. The elasticity helps to this purpose by conferring enough contact time for 'H' vectors from the respective rings to oppose; another contribution is facilitating the creation of nodes by the neutrons asymmetry, that makes them rotating structures, so they conveniently change the phase of its Hyle vector with respect to the proton vectors at contact point, allowing them to align. This mechanics explains why the neutrons that are intended to combine with protons in a nucleus should collide slowly.

## **Electron**

The electron is a type R- symmetric particle emitting Pp- photons as shown in **FIGURE XXXVII**. Even absorbing a low energy photon may cause its rupture when the neutron has enough energy and it arrives from the adequate direction, inducing additional shear.

Centrical electrons always receive low to medium energy photons from a nuclear proton; the intersection line XX of the electron discs and the proton discs will always be tangential since the discs from the emitted photon will always be in the same plane of the proton ring and in the same plane of the electron ring. This kind of absorption produces a tangential force generated by a highly concentrated group of nodes, compressing the electron perimeter, so prevent breaking apart while absorbing energy; this action results in a symmetric electron containing more energy than the standard free particle.

## Atomic nucleus formation

When a neutron increases its internal energy, it becomes asymmetric and its Hyle rotates faster; this condition facilitates node formation when colliding with a proton because, while the proton is a static structure the disordered orientation of the 'H' vectors in the neutron increases the probability of building a node by having two opposite vectors during the short period they are in contact. The duration of this contact is a function of their elasticity and the relative velocity; as mentioned before, this explains the need to slow neutrons in nuclear reactions in order to improve the probability in establishing a node. The next **Figure XLII** describes how protons and neutrons share nodes by knitting rings with diverse angles between their planes, for increasing the number of nested rings.

Once a node is formed, it must be strong enough to get both particles together, even when one of the rings gets disturbed by the absorption of medium energy photons.

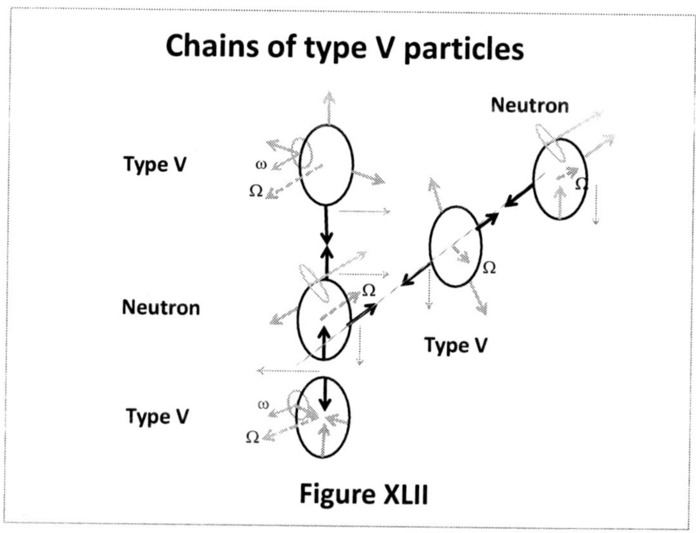



**Figure XLII**

To establish a stable node among the two particles, the neutron ring must rotate slower, therefore giving up some of its internal energy to the proton, which then increases its angular velocity '$\omega$' slightly reducing its diameter[123], finally both rings end up having the same size. The nuclear neutron anchored to the proton may constitute a new node with a new neutron-proton structure (nuclear fusion). This process liberates energy when the arriving nucleus has neutrons in excess, liberating them and/or emitting high energy photon; this analysis describes precisely the well know process of deuterium-tritium nuclear fusion.

The nuclear fission is the product of the weak bond of nodes in nucleus having too many protons, because the energy of each node in a ring decreases, as the quantity of nucleons around it increases.

---

[123] The node decreases proton period 'T*', increasing '$\omega$' and therefore increases the energy of the particle. The proton energy increase comes from the elastic energy of the neutron, which then decreases its asymmetry. Note that the maximum energy that could store a neutron is close to the energy of an electron; consequently, each one of the four nodes shared by a neutron with four nodes may have a "bonding energy" close to a multiple of 120 Mev per broken atom; part of it eventually released in a fission reaction.

## Nucleus geometry

The TH proposes that a nucleus is a spherical basket woven with proton rings and the neutron rings. They adopt a curved surface shape as a consequence of the geometric characteristics of the nodes shared by these two particle types.

When a stable particle like the proton (type V) builds a node with a similar size particle, type T like a neutron, the orientation of 'H' vectors at contact point enforces both rings structure a convex surface; for instance, a Helium nucleus as shown in **Figure XLIII**. The importance of having a convex surface is that the sub-space E+ planes (in dotted lines) generated by protons makes it possible having diverse, non parallel centrical planes which do not interfere each other, but may host two electron rings with similar diameter. Additionally it explains how only a proton with non occupied centrical, may send the compressed photons outside the atom, and then attract electrons.

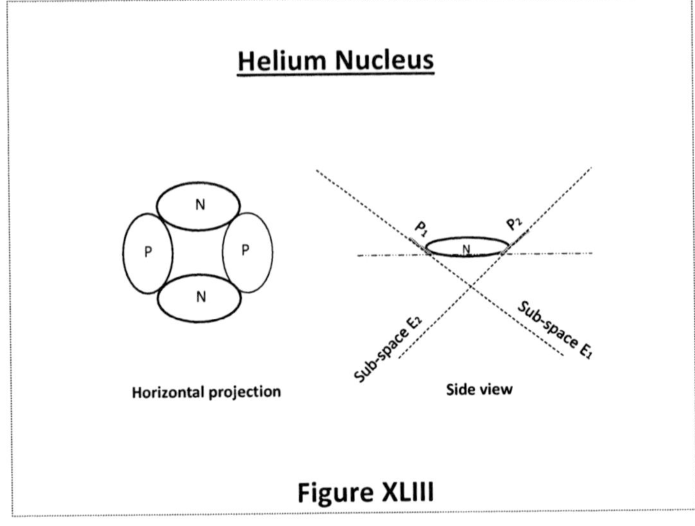



**Figure XLIII**

The quantity of nucleons (rings), shaping a spherical surface is defined by the atomic number[124]. Each ring may have 2, 3, 4, 5 or 6 nodes, as shown in the **Figures XLIV; XLV and XLVI** bellow:

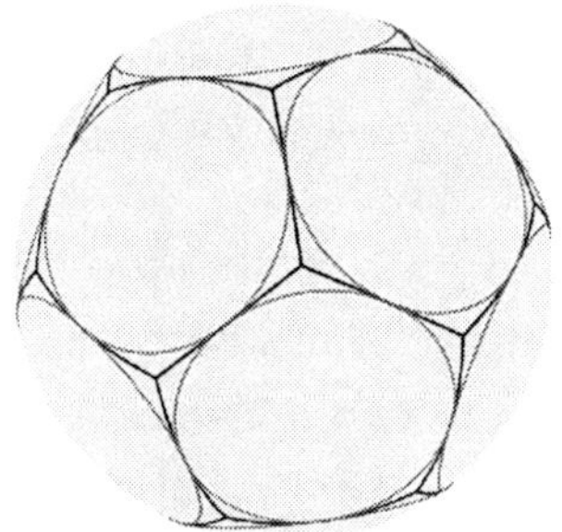

**Figure XLIV**

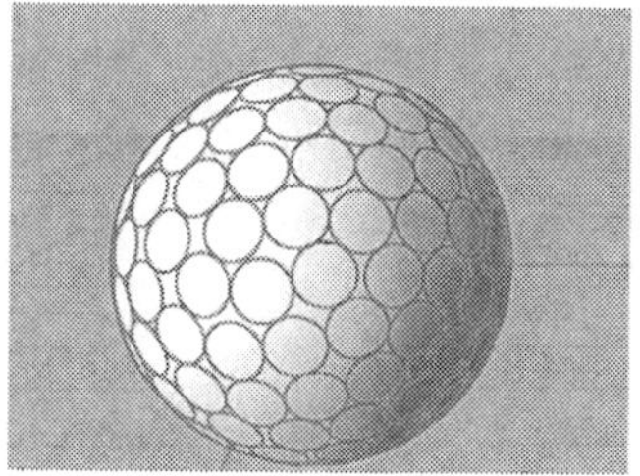

**Figure XLV**

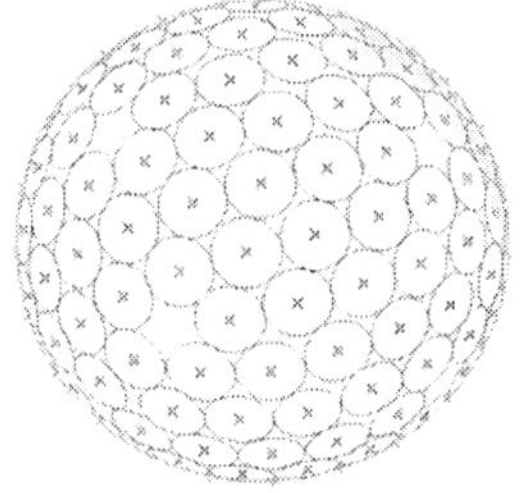

**Figure XLVI**

---

[124] The problem is similar to so-called "Tames problem" raised in biology when analyzing the number of corals in a colony.

The geometrical analysis on the shape of the spherical surface for the periodic table elements is a rather complex matter deserving the attention of future investigation. The so called "magic numbers", may result from a stability study on the spherically weaved surface of nuclear rings as a function of their quantity, the number of poles per ring, and how closed is the spherical surface they can weave.

A crucial characteristic of nuclear neutrons when they share a node with several protons is that all rings become static because the phase of the Hyle 'H' vectors from two particles at every node must remain constant at contact point, preventing them roll over each other. This objective is made possible by sharing enough energy in the node which is defined by the size of the sector of common perimeter; the proton absorbs a portion of energy from the neutron, reducing its diameter and making both rings change the angle of their ring planes, which was initially perpendicular. Therefore, as the length of the node grows, the angle between ring planes also increases.

The atoms from heavy elements include a large quantity of neutrons sharing nodes with a similar number of protons; when not disturbed, the whole structure is static. All the rings in the nest will rotate synchronously; when any of the nucleons absorbs a photon, transmits the rotation to all the rings in the structure (as when the drive gear rotates in a gear box); this mechanics effectively distributes, or better said, dilutes the asymmetry among all the rings in the nucleus; in this way the nucleus basket acts as a giant particle which absorbs photons and emits them homogeneously from several points on its surface[125].

---

[125] It is important to note that the energy difference between free protons and free neutrons is in the order of one in two thousands $[(\varepsilon_n - \varepsilon_p) / \varepsilon_n < 1 / 2000]$, which is also its difference in diameter; this energy is shared by neutrons and protons. Consequently, one neutron builds an energetic node with one proton; but each node will have a fraction of that energy when a neutron is linked to more protons around it.

This important nucleus characteristic explains how the nuclear metrix is added and why the emitted photons have an equivalent compression [126] equivalent to the one generated by a small ring having the total mass of the atom. Suppose that a proton or a neutron absorbs a photon, that particular ring increases its internal energy so it becomes asymmetric and rotates faster. As seen before, this rotation is transmitted to all the rings, so the whole structure will rotate synchronously; similarly, the whole structure stops when the nucleus emits energy surplus. It is reasonable assume that the nucleus absorbs several photons simultaneously and also emits the energy excess simultaneously; so the emission is permanent. Is overwhelming comprehending, how cleverly nature construct a giant structure that acts as a massive particle absorbs energy from random directions while emitting simultaneously photons in very well defined directions.

The nucleus of heavy elements is produced by a fusion process which demands bonding two nucleuses; not just adding an individual proton, because the increase of internal energy requires a source; and nature uses the surplus energy of nuclear neutrons which are destroyed or discarded in the process. When a new free proton is trapped by a nucleus, the diameter of all the rings in the nucleus is reduced. In conclusion, neutrons are reservoirs whose surplus energy may be used for multiple purposes.

The angular velocity '$\Omega_n$' of neutrons will point in opposite direction of the angular velocity '$\Omega_p$' of protons in the nucleus; so, assuming that

---

[126] When two static rings of similar size do share a node and one of them absorbs energy, it increases its angular velocity '$\omega$' while not changing its diameter; consequently it will start rotating and will transmit it to the next ring; but as both are material entities, they will behave exactly as a gear box with heavy gears, in a way that the rotational energy will propagate and distribute homogeneously. Therefore, when the atom resides in a sub-space location, all the rings in the nucleus will reduce their diameter in accordance to the "law of symmetry"; this phenomenon explains how the nucleus adopts the metrix of that position. A similar behavior is observed when a new neutron integrates to the nucleus; all the rings will slightly reduce their diameter, distributing proportionally the energy shared by the new particle.

neutrons have their '$\Omega_n$' pointing to the center, the '$\Omega_p$' of protons will point perpendicularly to the convex side the surface; towards the outside of the sphere.

Obviously a closed nuclear surface will have a spin of 1, because the angular momentum of a sphere is independent of the direction of the rotation axis. Most elements <u>do not have a closed spherical nucleus</u>; consequently the ratio of angular momentum for two perpendicular axis of rotation might be fractional.

## Nuclear emission and absorption

A photon arriving to the nucleus is absorbed by any of the particles of the basket of the static rings composing it. This absorption immediately generates asymmetry on the specific receiving ring forcing it to rotate and transmit this rotation immediately to all the 'A' nucleons (Z protons and N neutrons; A = Z + N) as if they were a single structure having a much higher inertia than any free nucleon; the angular momentum of the composite structure corresponds to the sum of the moments of all the nuclear rings in the nucleus; as they have the same diameter, all of them have the same inertia 'I'[127], so the inertia of the nucleus is (A I).

Considering that each proton has the same energy excess ($\Delta\varepsilon$), so we may assume that:

> a) Every nuclear proton emits simultaneously a photon because all protons have approximately the same properties, so they will break up simultaneously when reaching the same energy limit. Once a single energy packet is released, the

---

[127] All nucleus rings have the same diameter and the same rotation period; so they all have the same mass (m), diameter (D) and energy ($\varepsilon$). A detailed study of the absorption process will have to consider that any increase in angular speed implies that the elastic ring perimeter slowly stretches, adding a delay to the transmission of the rotational momentum to other rings.

emission process will continue until the Hyle structure extracts '$\Delta\varepsilon$' which defines the duration of the emission that is the photon period '$T_o$'.

b) Every photon emitted by a proton in the nucleus has the same energy ($\Delta\varepsilon$).

Consequently, the total energy released by the whole basket should be (A $\Delta\varepsilon$) and will be emitted by every one of the Z protons in the nucleus; therefore, every photon will have a period ($T_o{}^* = h \, Z \, / \, A \, \Delta\varepsilon$); in most cases (Z = A / 2), then ($T_o{}^* = h \, / \, 2 \, \Delta\varepsilon$). Therefore, the emitted Pp+ has:

A) A compression ($\xi = A \, \Delta T_o \, / \, T_o$) approximately proportional to the total mass of nucleons.
B) Information that the emitter is a particle with positive charge; this characteristic means that its disc always rests on the same plane than proton ring. This is the key geometric property which structures centricals.

The compression '$\xi$' of a photon with energy '$\varepsilon_o$' emitted by a free particle is independent of the released energy ($\Delta\varepsilon_1 = \varepsilon_o$); the compression is only proportional to the ring curvature '$\kappa$'; then ($\xi = k \, \kappa$). The curvature of a particle ring is proportional to the emitter mass ($\kappa = k \, m_1$), and the mass is inversely proportional to the period ($m_1 = k \, / \, T_1{}^*$; or, $T_1{}^* = k \, / \, m_1 = k \, / \, \kappa$). Taking into account that most emitters are practically symmetrical [($T_1{}^* - T_1$) / $T_1$ < 0.01]; consequently, then the relative photon compression ($\xi = \Delta T_o \, / \, T_o = \Delta T_o \, / \, T_o{}^*$) is a dimensionless parameter virtually independent of the emitted energy.

All "A" nucleons have the same perimeter '$T_1$', the same mass '$m_1$', and the same curvature '$\kappa_1$'; therefore, an hypothetical free particle with mass '$m_2$' equivalent to the nucleus mass ($m_2 = A \, m_1$), would be a ring with a very small period '$T_2{}^*$', and a perimeter close to ($T_2 = T_2{}^* =$

$T_1{}^*$ / A). This hypothetical particle would emit photons with a compression ($\xi = \Delta T_o$ / $T_o$) inversely proportional to '$T_2{}^*$' or proportional to its curvature '$\kappa_2$' ($\xi = k$ / $T_2{}^* = = k\,\kappa_2 = k\,A$ / $T1^*$). The nucleus emits a similar photon because it has an equivalent inertia[128].

It has been shown that the nucleus reacts as a single giant Hyle structure because any change in rotation speed induced by absorption is instantaneously transmitted to all rings, which strongly resist any angular acceleration because of its individual inertia; consequently, the emission process will be comparable to the emission from a very small ring with an inertial equivalent to the total inertia of nuclear rings.

If one considers that the emission of photons peels off a thin energy layer from a fast rotating small ring, it may be as peeling off, the same thickness of energy from a slowly rotating large structure with equivalent inertia[129].

Comparing the total mass (energy) of a nucleus with the addition of individual masses from the free nucleons it contains, should always consider the following facts:

---

[128] A material ring, boosted by a tangential force changes its angular velocity; in terms of energy, the increase in rotational energy is given by the  equation $[\Delta\varepsilon = (\Delta I\,\omega^2 + 2\,I\,\omega\,\Delta\omega)\,/\,2 = \Delta I\,\omega^2\,/\,2 + I\,\omega\,\Delta\omega]$; i.e. when the increase in inertia is negligible then, $(\Delta\varepsilon = I\,\omega\,\Delta\omega = L\,\Delta\omega)$, so the angular momentum 'L' of the particle determines the amount of energy required to change its rotation period; therefore, the increase of ring energy <u>is proportional to the increase of the apparent angular velocity</u>. Therefore, the energy of a nucleus increases when the angular velocity of the train of rings increases; as the individual inertia is 'I', the total inertia is $[A\,I = (Z + N)\,I]$. This explanation is also coherent with the emission model that establishes that the compression is proportional to the curvature which in turn is proportional to the inertia.

[129] The thin layer of energy taken from a small diameter ring may take thousands of revolutions while the equivalent thin layer taken from a large structure with the same mass rotating much slower will demand only few revolutions; the process in both case could be compared to unwind a fine cord of the same "length" from a fast rotating small reel or from a slowly spinning large bobbin.

I.- The actual energy of a free neutron is a statistical parameter whose magnitude is the average energy of many neutrons, each one having more energy than the basic energy level '$\varepsilon_n$' of a symmetric neutron.

II.- A fraction '$\alpha$' of the surplus energy '$\Delta\varepsilon$' of a neutron in a nucleus is absorbed by the proton when building a node; the node increases proton energy by ($\alpha\,\Delta\varepsilon$) and decreases neutron energy by the same amount.

III.- The remaining energy ($\alpha\,\Delta\varepsilon - \Delta\varepsilon$) will be released as a photon during the binding process.

IV.- In general, the nuclear transmutation process is much more complex than the simple description of a neutron colliding at right speed and direction with an already existing nucleus. The process demands precise manipulation recipes, normally involving other nucleus which bonds some its particles with the original nucleus, and releases unused particles.

It is important note that the TH proposal on nuclear mechanics explains important aspects of the atomic phenomena, as:

- Explains how the nuclear structure emits photons with a compression proportional to the total mass of the emitting body, caused by the "basket like" nucleus structure which reacts as a giant single particle.

- Explains why the net charge of a complete atom is zero for an external observer, and how the positive charge of the nucleus emerges externally only when there is no centrical electron, because it prevents the Pp+ emitted by a proton could generates a sub-space $E^+$, by absorbing it previously.

- Explains why the magnitude of the "electric field" (sub-space E) is proportional to the total charge of the nucleus (double of it), and how this "field" automatically aligns a free electron for occupying that centrical position.

- Explains why there are nuclear structures that show more stability than others (magic numbers), as the consequence of, how closed the spherical nucleus structure is.

- Explains why the electrons in an atom always reside on a particular proton ring plane, even when changes energy level; because the nuclear proton ring always traps electrons on its own plane.

- Explains how protons and neutrons get linked together, stabilizing the neutron.

- Explains the detailed nuclear mechanics, in order to store and emit energy, unveiling the forces involved in this process.

- Explains why the spin of most nucleuses is close to 1; because a nucleus is seldom a perfectly closed sphere.

**Gravitational and electric fields ratio**

The conventional description on the electric and gravitational interaction between large groups of individual particles using the abstract concept of "field" could derive in imprecision, or some improper description about these two supposedly isolated phenomena, particularly in the case of fields generated by a nucleus, as explained in the following lines.

Nature always establishes a field of velocities 'v' around any material body, also known as "gravity potential." The integral of this three-dimensional "velocity field" results in the three dimensions of $\pi$

conventionally known as "space" where the distance on a coordinate is ($z = \Sigma \, v \, \Delta\tau_z$). The associated acceleration field, conventionally known as "gravitaty field", may be found by deriving the velocity ($a = dv/d\tau$). The magnitude of the velocity in a "velocity field" at a distance 'z' is proportional to the mass 'm' of the material object which creates it, and it is inversely proportional to its distance, so the magnitude of the acceleration <u>at a location is also proportional to the emitter mass.</u>

The "distance unit" close to the emitter appears shorter when the velocity field is intense, because a moving object appears to an external observer as travelling slower, in this way we may say that distances are shortened by a compression factor; so does the unit of distance given by the fractional metrix comparing distances at sub-space G in reference to an ideal flat space distances.

A velocity field <u>generated by an arbitrarily defined point of unitary mass,</u> is a universal geometric constant (its spherical distribution is inversely proportional to the diameter of the sphere); we will call it "standard velocity field" from which we can find a *standard gravitational field*" by deriving it. This means that any unit mass placed at a certain distance from an emitter of unit mass will also travel a constant distance '$d_o$' per time unit so will acquire a standard acceleration '$g_o$'.

In the case of an "electric field" is improper describe it as a "velocity field", because it does not establish directly the speed of the receiver in spatial distance units and universal time unit; the velocity depends on the mass of the receiver charge. Consequently, will be also improper refer to it as an "acceleration field", because the acceleration will also be function of the mass; consequently, it is more accurate refer to it as an "electric force field"; so, <u>electric fields, are force field residing in a gravitational "space."</u>

Experience indicates that the electric field may produce much higher forces than any gravitational field, so we can say that the electric force applied to a unit charge with unitary mass ($m = 1$) in a "standard electric field" accelerates it at ($a_e = F_e / m$) which is larger ($a_e > a_g$) than the "standard acceleration" 'g' of the particle in a gravity field ($a_g = F_g / m$).

A general electric force field cannot be described an acceleration or velocity field, because these terms imply referring to an unknown mass containing the charged emitter. The electric force is inversely proportional to the square of the distance; parameter whose magnitude is always established by the sub-space G as universal position reference for any particle; charge or not.

As "space" is an abstract entity originated from a natural velocity field, the variable "distance" for both fields is set by the travel time of a photon; but we must remember that is the same photon which carries the electrical and gravitational information; consequently, the electrical quality (sign and charge) is a binary attribute of a charge, specified by a geometric property to the Pp it emits, so the geometric interaction during absorption, applies larger (electric) forces on a "charged" receiver. Comparing Coulomb's law ($F_e / q^2 = k_e / r^2$) with Newton's law ($F_g / M\, m = G / r_2$) in the case of standard fields, ($M = m = q = 1$), we find that its ratio is a constant 'η' ($F_e / F_g = k_e / G = \eta$) representing a "multiplier force factor" 'η' characterizing electric force field over gravitational force fields.

It has been shown previously that the ring of a free charge absorbing a Pp, is subject to a tangential force which impulses it roll moving its center a longer distance than the same particle moves when subject to the random forces generated by Np nodes.

A massive emitter, as is the atom nucleus, sets the metric of the surrounding sub-nuclear space; the emitted Pp+ indicates that a charged particle generated it. Note that the term "charge" relates to the magnitude 'q', which only indicates the geometry of emitter; it is a binary variable without physical dimension. A single Pp just indicates that the emitter has a given polarity; it does not carry the magnitude of the applied force; this force depends on the geometry of the receiver and the intensity of the gravitational field. This analysis clarifies that electric fields can be added <u>only as cumulative sequential forces</u> applied by every photon absorbed by one or many receivers sharing a material structure.

A charged emitter always has a unitary charge ($q = 1$) and so does the receiver; then, the electric attraction force given by the coulomb's law is inversely proportional to the distance between the two particles ($F_e = k_e / r^2$), and distance is a natural parameter resultant from the TT of a compressed photon advancing one length per revolution. Newton's gravitation law establishes that the gravitational force produced by a unit mass emitter to a unit mass receiver is ($F_g = G / r^2$). As already explained, the geometry of the Pp generates an electric force ($F_e = \eta F_g$) applied to a unit charge positioned at the same distance, where '$\eta$' is a dimensionless universal magnification factor ($\eta = k_e / G$) generated by the tangential orientation of forces from all nodes produced by a Pp on a "charged" particle. The nucleus-electron interaction within atoms is extremely finely tuned as shown by the following numbers: the diameter of an electron is 1800 times the diameter of a proton, and the emitted photon may be $10^9$ time larger. If we set each energy packet has 0.1 eV, the proton will have $10^{10}$ packets, the electron 5 $10^6$, and the photon $10^3$ packets; so the proton sends one packet by every $10^7$ revolutions.

The gravity field strength '$F_g$' in the vicinity of an atom is proportional to the nucleus mass, including all 'A' nucleons, each with an individual mass 'm'; therefore the electric force '$F_e$' is '$\eta$' times larger than '$F_g$', because it is the consequence of the photon architecture notifying that the emitter total mass 'M' is ($M = A\,m$), and that it was generated by a charged particle, so it will produce an electrical attraction equivalent to that it would have a single proton with a total charge ($Q = A\,q$), as if the nucleus would have 'A' protons; therefore, the nucleus creates an electric field twice as large as the one generated by the sum of individual protons[130].

A notion that should be emphasized because of its conceptual importance is: there is no "space" when there is no a stable gravitational field; a flat space is an imaginary entity. Consequently, there cannot be an electric field without an underlying three-dimensional space; of course it may be an isolated Pp arriving to a receiver, but an occasional signal does not generate a "field."

Finally it is important standing out the dangers on using classical perspective and conventional semantics in describing the dependence between electric and gravitational phenomena, since it may originate contradictory concepts, fruit of entelechies which oversimplify the complex interaction of individual particles.

Distance is the basic parameter that determines the magnitude of electric forces between two charges, and this "distance" is necessarily defined by a sub-space G (gravitational field); then, the magnitude of the "field E" depends on the magnitude of the field G, which means that the magnitude of the electric force of attraction/repulsion in a "standard electric field" is always proportional to the gravitational force in a "standard gravity field." Additionally, this statement suggests the mechanics of action-reaction rules also in electric phenomena.

The electric phenomenon could be properly expressed at individual level by saying:

A Pp emitted by a charged particle contained in a massive body, has two characteristics: a) its compression is proportional to total mass of the emitter, and b) its structural type defines the polarity of the emitting charge; in other words, one gives gravity information, and the

---

[130] This is a phenomenon demanding further analysis.

second electric information. Then it is absolutely clear that <u>the same carrier communicates simultaneously information on two independent physical parameters.</u> Although considering that they cannot be communicated separately, the electric force applied to the charged receiver will be proportional to the gravity force (compression) multiplied by 'η'; therefore, when a photon communicates electric force is communicating simultaneously its compression, so also defines the gravity acceleration.

Another concept that should always be accounted for, is that the metrix of a sub-space G is adopted by all particles <u>residing stationary</u> [131] <u>in a spatial location</u>; consequently, when they absorb a compressed photon (Np or Pp), they do not move because the particle emits a photon containing equivalent energy and the same compression, than the one received.

## Nucleus and atom metrix

The TH defines metrix as the ratio of the increase of TT 'τ' per absolute time unit ($\mu = \tau / t$). The most basic Hyle open structure is the photon which defines the sub-space metrix by increasing its distance to the emitter by its "length" '$T_o$' per revolution period '$T_o*$' so ($\mu = T_o / T_o*$). This dimensionless parameter cannot be applied to a particle as the ratio of its perimeter '$T_1$' and its period '$T_1*$', since the ratio ($T_1 / T_1*$) will always be one because a particle that does not move in a sub-space must always be a symmetric ring which appears static to any external observer. Considering that the compression '$\xi$' of the photon emitted by a particle is given by the ratio of its length reduction '$\Delta T_o$' and its length ($T_o = T_o*$); so ($\xi = \Delta T_o / T_o*$), and as previously demonstrated, '$\xi$' is proportional to the mass 'm', or to the curvature 'k'; therefore, <u>the extra compression</u> of photons emitted by a particle

---

[131] The metrix of a ring always refers to the ratio ($\varepsilon_1 / \varepsilon_m = T_m / T_1$) where '$\varepsilon_m$' is the nominal energy level for that particle type and '$\varepsilon_1$' its actual energy.

with fractional metrix will be defined by the decrease of its perimeter '$T_1$' (smaller diameter) when it contains the same amount of released energy; consequently, the metrix of a particle could be defined as the ratio of actual diameter and nominal diameter ($\mu = T_1 / T_n$), and a "compressed particle" may be defined as the one that has a smaller diameter than nominal.

It has been shown previously that there may be symmetric particles which store more energy; the symmetry of a free particle which stores some extra energy can be induced by external influences as the nucleus architecture, which generates internal compression while the particle absorbs energy, forces [132] it keeping its perimeter smaller than nominal.

The atomic nucleus as a whole, acts as a free particle by adopting the metrix of the surrounding space when it does not change relative positions with neighboring atoms. It absorbs photons that increase its internal energy while the perimeter of all their rings is slowly reduced to the same metrix of the arriving photons because of the permanent effect of internal elasticity of its rings. This cumulative process allows atoms share metrix with the surrounding environment by sequentially increasing nucleus energy until it achieves symmetry while reducing the diameter of its rings. The nucleus of an atom which is part of a large material object will emit Pp+ whose compression is defined by its internal metrix defined by the total mass of all atoms in the object.

The atomic electrons do not contribute significantly to the nuclear metrix because of their relatively small energy content, and they communicate their properties to their own atom only indirectly,

---

[132] There are cases when the energy absorption process itself compresses the ring perimeter, as in the case of the nucleus or the centrical electrons; other alternative to compress a ring is to apply an external pressure, as observed in objects resting on a solid that is in a permanent position in a sub-space.

because orbiting electrons always emit photons tangentially to the external sub-space, <u>always travelling away from the nucleus</u>.

## Electrons in atoms

As previously explained, the position of a circular ring around a center is not comparable to a planetary orbit as initially proposed by early atomic models, neither is an abstract energy level occupied by a probabilistic cloud as seen by the quantum mechanics mathematical models. The TH adopts the name of "centrical", as the generic denomination of the <u>electron ring position</u> around the nucleus, each one having an electron with specific geometric and physical properties; this denomination was adopted because it evokes both concepts; energy level content as in Quantum Mechanics, resembling also a circular path (orbit) around the nucleus, as in classical mechanics.

A free electron is attracted by a nucleus whenever there is an unoccupied centrical; which physically is represented by the plane of a sub-space belonging to a specific proton. When the electron enters to the strong sub-space E generated by the highly compressed Pp+ emitted by this proton, the ring is subject to a strong attraction that will make it roll towards the emitter, simultaneously aligning its plane with the subspace plane [133] which finally places it with the nucleus in its center. Any vertical or radial force applied to the approaching electron ring, makes it rotate on its diameter (perpendicular to the photon path), cleverly causing a precession which rotates the ring diameter placing it on the same plane of the proton ring.

---

[133] The free electron approaching process should be very complex because initially the ring plane angle with the sub-space E may be anyone, but the geometric characteristics of the intersection line XX and the plane of the disc in a type R particle will produce many effects: rotate the ring, align its plane, bring it closer to the nucleus, place it concentrically, and finally compress it until the ring acquires a static symmetric structure storing the energy surplus relative to that centrical.

The sub-space E⁺ generated by protons in the nucleus, attracts any free electron which crosses it, placing it on the same plane than the proton ring plane. When the nucleus is in its center, the composite geometry of the rings, discs and sub-spaces, defines that the intersection line XX of the photon discs with the electron ring be always tangential to the electron ring as shown in **Figures XLVII** and **XLVIII**; considering that the photon disc plane is always on the same plane of the proton ring, photon discs will also be on the same plane, defining a <u>stable tangential intersection line XX during absorption</u>.

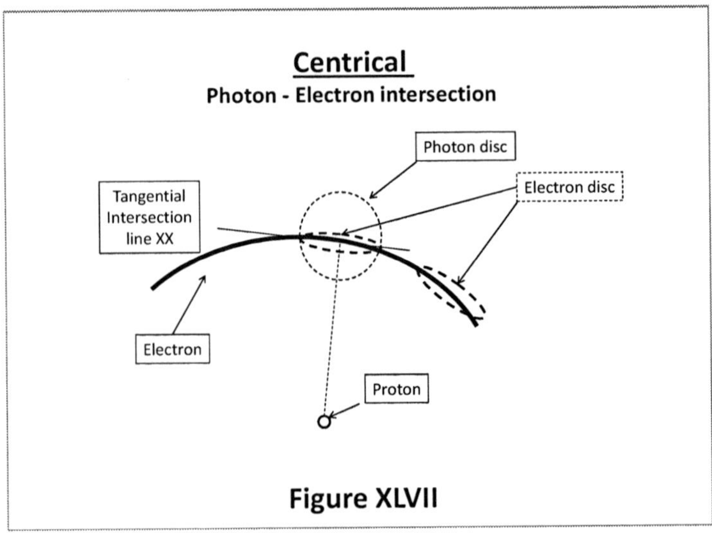



**Figure XLVII**

This unique property of the three particles involved in the process conveniently produces a very strong tangential elastic compression on the electron perimeter; so it does not produce shear tensions which may break the ring. In this manner, the gap generated in the ring will be closed by the very strong elastic force aligned with the circular own space of the electron; so achieving two objectives; a) that the energy level be proportional to the nuclear mass, and b) that the plane of the electron remains aligned because all the photons absorbed will always prevent any deviation.

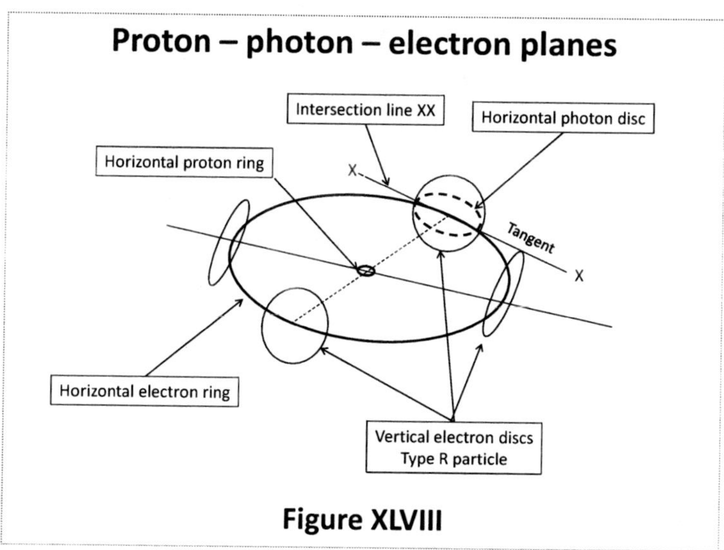



**Figure XLVIII**

The highly compressed photon emitted by the massive nucleus does not have the time for decompressing before reaching the centrical because it is hundreds of thousands times larger than the diameter of the electron ring, so the gravitational sub-space G _permanently compresses the electron ring_ proportionally to the nucleus total mass, while prevents electron ring tilts out of the sub-space E$^+$.

Considering that every proton in the spherical nucleus has its sub-space plane oriented in the tangential plane of the sphere, all the sub-space E+ planes, have to point to different directions; so, there will never be a nucleus with pair number of protons in any equatorial circle.

**<u>Energy levels</u>**

It has been shown that each electron occupies always the plane established by its master proton on a centrical, so the stored energy by it may be any, it could be the highest or the lowest; the maximum energy "level" will be defined by the nucleus mass and its particular position by the available centricals. Consequently, the geometric location of an electron is independent [134] of the energy it stores; any atomic electron may change diameter; from having the smallest diameter set by the particular nucleus, or have the largest diameter when all other energy levels are occupied. The permanent tangential compression of the centrical electrons will drive all of them to occupy the maximum energy level available.

It is also important observe that the compression applied to any atomic electron by any Pp+ emitted by the nucleus is always approximately the same, because its compression is defined only by the nuclear mass. Consequently, every centrical electron absorbs enough energy in order to reduce its diameter to the minimum allowed dimension for reaching symmetry by matching minimum diameter with maximum energy level. The result will be a ring with the largest attainable energy level to have a symmetric ring ($T^* = T$) having the metrix of the sub-space where it resides; in conventional language we may say that it reaches the energy level given by the Eigen values for Schrodinger equations.

When a photon emitted by the nucleus is absorbed by a centrical, the electron internal energy increases by decreasing its period '$T_1^*$' until it attains symmetry. The first electron trapped by the nucleus will adopt the minimum diameter '$T_1$' defined by the nucleus mass, since the ring diameter is fixed by the compression of the internal photons emitted

---

[134] This may be the reason of the vague positioning (clouds) found by Quantum Mechanics mathematical analysis. In general, a particle emits a photon and migrates from high to low energy levels, increases its diameter, so in the case of an centrical electron it will find a net of rings blocking its way out, so it is reasonable to assume that the atomic mechanics will follow a "push-up" sequential protocol.

by a proton in the nucleus. It will not absorb photons anymore because its diameter '$T_1$' compared with the nominal diameter '$T_n$' of a free electron ($T_1 / T_n < 1$) will have the same metrix than the next arriving photons and therefore the same metrix of the nucleus. This electron may change its energy level by emitting a photon with a compression equivalent to its curvature, containing the amount of energy set by the decrease in its energy level.

## Centrical pairs

The perimeter of an electron ring "B" which is in the process of reducing its diameter by absorbing an "internal" Pp+ emitted by the nucleus, will meet another stable ring "A" already occupying that energy level but in another plane; considering that "B" is asymmetric, and "A" is not, it will be ease for both rings establish a pair of opposite nodes on a common diameter located in the intersection line between both rings. In this way, finally both rings will adopt the same diameter and will share the same energy level, as shown in the next **Figure XLIX**. Any extra energy of the arriving electron will be emitted, so both electrons sharing the same energy level will be symmetric structures.

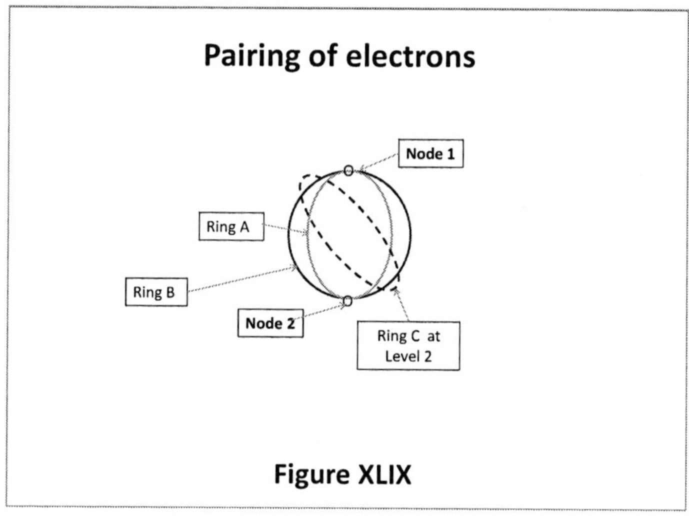


**Figure XLIX**

When a third electron ring "C" reduces its diameter by absorbing a Pp+ from the nucleus, will occupy the same energy level already occupied by "A" and "B"; obviously, due to a phase difference, it cannot establish simultaneously nodes with rings "A" and "B", so it will bounce back to the next lower energy state in which it may achieve symmetry (conventionally found by the Eigen states given by the time independent Schrödinger equation), whatever energy excess will be released as a photon.

Once the electron ring reaches a stable position, it will not absorb more Pp+, returning them, back to the nucleus; or if it absorbs them, it will emit an equivalent amount of energy. This procedure explains the sequential stable occupation of energy levels and why there are two and only two electrons in the same centrical sharing the same energy level; it is important note also that whatever is the energy level, the plane of the centrical is always defined by the direction of the plane of a proton in the nucleus.

At this point, I cannot resist saying that the process is astounding, by the cleverness and simplicity by with which nature automatically builds complex atomic structures, using such simple and orderly process.

## External energy absorption

When an external photon reaches an electron centrical; whatever its type, direction of arrival, or energy content, produces shear forces which may fracture the ring so its energy will increase immediately while its diameter will start contracting very slowly. Let's suppose that ring "A" increases its energy substantially, destabilizing ring symmetry breaking the nodes shared with "B"; the ring "A" will become highly asymmetric, probably even rotating the plane of its discs; the only way for returning to a symmetric type R structure (electron), is by releasing the tangential (and eventually vertical) Hyle vector component,[135] by emitting a Np.

## Schrödinger and Hyle equations

The Hyle equation "H($\tau$; t)" <u>describes the substance constituting any particle</u> as a three dimensional periodic vector function, with magnitude equal to one within the limits of its structure; it also shows that material particles are circular static entities while photons are linear structures travelling away from its source while rotating on a two-dimension plane. By following the most fundamental natural law, the "Law of Symmetry", the TH shows how a geometric analysis of this proposal explains the origin of the basic variables in physics and most of the observed natural phenomena.

The Schrödinger "Wave Function" "$\Psi$(r; t)" describes particles as oscillatory abstract entities in accordance to the paradigm of the "principle of wave-matter duality", which is based in the conjecture proposed by De Broglie, that a particle could be described as an oscillating energy.

Both equations may look similar because they include the same variables[136], and deal with the same object basically described by the Einstein-Planck equation relating angular speed (time) with energy. But there is a substantial difference: the TH is a proposal which describes how universe is constructed under a unique perspective,

---

[135] Normally, the emitted photon will have its discs plane perpendicular to the tangent, in this way the emitted photon will contain all non tangential components of the Hyle vector, so this photon will be a NP that communicates that the atom has no charge. These Np are the most common in an environment with many atoms. It is also possible that some external photon collides with the centrical electron and change the direction of the ring plane; freeing it from that centrical plane, so releasing the electron from the atom.

[136] The variable 'r' in the wave function represents a three dimensional vector location at time 't', which is not a universal variable under the postulates of the "Theory of Relativity." The variable '$\tau$' in the Hyle equation is the unit vector pointing to a direction in a three dimensional "travel time" environment, and time variable 't' is the absolute universal time.

under which energy is a secondary parameter derived from the effects of disturbing the elastic condition of Hyle structures; while the Hamiltonians and Laplacians used by the Schrodinger equations are only equations applied to Newtonian mechanics definitions, not contributing with any substantial explanation on origin of this natural phenomenon.

From a mathematical point of view, the Hyle equation is equivalent to "the time-dependent three dimensional Schrödinger Wave Function"; with a fundamental difference, the "wave function" is an abstract mathematic entity based of the De Broglie conjecture, while the "Hyle equation" is the consequence about the necessary oscillatory condition of an essential substance conforming everything. The TH proposes and demonstrates that a particle is not a dual abstract entity, it is a substantial entity always occupying a location relative to other natural entities; it has geometrical and physical properties, and it interacts ruled by precise geometric theorems.

The standing wave proposed by Schrodinger, could be associated to the way the TH describes the symmetric condition of any stable particle as a <u>static and unchanging three dimensional entity;</u> similar to, but not a standing wave.

The mathematical tools and procedures followed by Schrodinger are valid and could be applied to Hyle structures, and obtain similar results. From a mathematic perspective, the differences between both visions may appear be merely formal, but there is a profound difference from the physics point of view, because in the TH the proposed entity is engendered as the substantial building block ruled by a single law "the law of symmetry" which explains coherently diverse phenomena as gravity, force, time, etc., "rediscovering" fundamental physics laws/equations and finally explaining with the same components the complex atomic mechanics; on the other hand, the "wave function" describes a vague mathematical entity with uncertain geometry and undetermined location; all we know about the particle, is that contains oscillating energy.

The Schrödinger equations, the Eigen states, or any other appropriate mathematical procedure could be applied in order to find the centrical energy by accommodating the Hyle equations to the TH description. Moreover it will be simpler and more illustrative for a physicist apply mathematic procedures to a model that describes the real shape of the investigated object, its physical properties (energy, momentum, elasticity), and most important of all, that it has close ties with basic concepts as natural causality, contingent quality of space and absolute time; rather than working blindly, as a pure mathematician, with abstract ideal objects. However, it is predictable that the any change on investigation methodology, ignoring paradigms, will demand a deep cultural adaptation.

**<u>This page is intentionally left in blank</u>**

## Chapter XVII

## Cosmology

This chapter analyzes phenomena involving long distance material interaction; also offers explanations to diverse subjects, under the TH perspective, including the analysis of "principles", which currently do not explain the origins of phenomena[137]. Some of them are: the expansion of the universe; the size of galaxies; material objects moving faster than light speed; the gravity generated by massive objects; and finally a proposal on the birth of the universe.

### The Hubble effect

The space metrix ($\mu = d\tau/dt$) at a point in space, is established by the compression of photons ($\mu = T_o / T_o^*$) passing through that point; additionally one should remember that the initial photon compression is proportional to the emitter mass, and that the compression of any photon decreases as travel time increases; this phenomenon is conventionally described as a curved space with convex metrix generated by a decreasing velocity field which is the origin of gravity; any object on it, moves at the velocity set by the average photon compression in that location. Properly speaking, <u>the space is the entity that moves towards the emitter,</u> faster as it gets closer to it.

Considering that the photon compression decreases asymptotically, it becomes negligible after a long TT; consequently the space becomes practically flat, so the position of an object positioned at long enough distance from the emitter will not change when it receives a signal

---

[137] During the development of TH proposal, I systematically avoid referring to "principles", which generically evoke "unexplained natural phenomena." Normally a "principle" refers to an observed natural phenomenon which is possible to describe it mathematically, but whose causes are unknown.

from it; but experimental observation of distant stars and galaxies show that they recede faster as their distance increases; this relation is linear, and its slope is established by the Hubble constant "Hu"[138], which defines the perceived rate of speed increase as the emitter increases distance.

Under the TH perspective the Hubble effect can only be explained as a consequence of the permanent tiny increase in photon wavelength which becomes relevant after very long travel-times. The Hubble constant establishes the relative acceleration perceived by a receiver when the remote emitter recedes faster as it gets farther; the increase in distance is not generated by the emitter moving away but by the expansion of photon metrix; in conventional language we may say that the space is expanding. The local gravitational metrix around the receiver is not modified by this effect; the space contains many more photons from close emitters than some few old photons arriving from distant objects. In accordance with this prognosis, the aging of photons produces a perpetual increase of distance between galaxies.

From the particle's perspective, photons are open structures of Hyle that do not benefit from the stabilizing feedback effect that particles have because Hyle is moving on a circular closed path of their own space; their circular geometry offers the mechanics to permanently self correct any divergence between spatial and time phase of Hyle. From this point of view, the difference in path geometry from a straight line to a ring has a much deeper meaning than just a geometric shape; from being a delicate linear information carrier, it becomes a stable entity instituting permanent structures of rigid matter.

---

[138] Hu is approximately 70 (Km/sec)/Mpc, and 1Mpc is $3\ 10^{19}$ Km; then Hu = $23\ 10^{-19}$ /sec. One Light-year in TT units is $2.3\ 10^{6}$ sec.

A photon undergoing a structural change in terms of its relative scale has only one possible explanation: that its internal length '$T_o$' growths gradually as AT '$t$', or TT '$\tau$' increases.

Let's analyze a simple model assuming that relative period $(\Delta T / T)$ increases linearly with travel-time. "$\tau$"; this is:

$$\Delta T / T = \sigma \tau \qquad (17\text{-}1)$$

Considering that photon compression $(\xi = \Delta T_o / T_o = T_o{}^*/ T_o - 1)$, then '$\sigma$' defines growth rate of spatial metrix in relation to travel-time. In accordance with Doppler relation, the apparent speed perceived by the receiver is $(v_h = c\, \Delta T / T)$; then:

$$v_h = c\, \sigma\, \tau \qquad (17\text{-}2)$$

$$\sigma\, \tau = v_h / c \qquad (17\text{-}3)$$

Or: $$\sigma = (v_h / c)/ \tau \qquad (17\text{-}4)$$

So $(1/\sigma)$ is a large constant in time units. The apparent velocity will be the light speed '$c$' when the travelled time is $(\tau = 1 / \sigma)$.

Considering that the photon expansion has no limit, there may be extremely distant objects travelling away at an apparent speed faster than light speed '$c$'[139].

---

[139] A photon always increases permanently its distance relative to its emitter, it is important to note that the at the beginning a photon travelled away from its source (the distant galaxy) a smaller distance than it should increase if travelling in a flat space; and after very long time it will travel away faster. This is not easy to understand under the conventional perspective that gives to "space" a physical meaning; under this erroneous point of view, one cannot explain the origin of the "energy" needed to push away distant objects, and how the local space is convex while the distant "space" is concave. For the TH is very easy to explain that the meaning of "distance" or "distance increase" of an object may be independent of the metrix of local "space", for self propelled objects.

The negative acceleration '$a_h$' for intergalactic "space dilatation" is then a constant:

$$a_h = dv_h/d\tau = \sigma\, c \qquad (17\text{-}5)$$

This acceleration is given by the Hubble constant 'Hu' as:

$$Hu = \Delta v_h / d = \Delta v_h / c\, \Delta\tau = a_h / c \qquad (17\text{-}6)$$

Then: $\qquad\qquad\qquad a_h = c\, Hu \qquad\qquad\qquad (17\text{-}7)$

The concept of Hubble's constant [140] being a negative acceleration (opposite to gravity) explains how nature limits the size of material conglomerates, as the galaxies.

The acceleration is:

$$a_h = dv_h/d\tau = \sigma\, c = c\, Hu \qquad (17\text{-}8)$$

Then $\qquad\qquad\qquad \sigma = Hu \qquad\qquad\qquad (17\text{-}9)$

### Complete Hyle expansion equation

Nature dictates that universe size increases permanently, isolating matter[141], so effectively universe ultimate fate is eternal separation of galaxies; which will get cooler until the lack of communication separates their particles.

---

[140] Considering that the value of '$\sigma$' is ($\sigma = 23 \times 10^{-19}$ [1/sec]), then the travel time distance for a galaxy to recede at light speed is [$4.3 \times 10^{17}$ sec = $1.3 \times 10^{10}$ (light) years = 13,000 million years in TT units].

[141] The metrix of few photons arriving from a distant galaxy do not establish a local "space", neither a "local velocity field" which meaningfully affects the receiver relative position; these few photons only inform that the emitter is receding at a given speed. The intergalactic "space" is not really a "space"; it is the addition of the very weak effects from few photons emitted by very distant objects that do not modify the relative position of the receiver. When a self propelled object travels in a "weak" space, the term "distance" and "velocity" should be more properly referred to parameters related to its own environment, not to the signals from distant emitters which are only informative, not causing any measurable effect.

The Hubble effect indicates that old photons have a metrix larger than one ($\mu > 1$); consequently, the Hyle equation must be completed in order to include this term. Remember that the asymptotic photon decompression relation with TT is ($\xi = \Delta T / T = A_g / \tau$), so it must include a linear expansion term ($+\sigma\,\tau$) that will permanently increase its length at a very small rate; the effect of this expansion is unnoticeable for distances smaller than 100.000 light-years. The complete formula ruling the length of a photon becomes:

$$\xi = \Delta T_o / T_o = A_g / \tau + \sigma\,\tau \qquad (17\text{-}10)$$

Since: 
$$A_g = G\,M / c^3 \qquad (17\text{-}11)$$

Then: 
$$\xi = G\,M / c^3\,\tau + Hu\,\tau \qquad (17\text{-}12)$$

The "distance" from the receiver to the emitter is a contingent parameter defined by the travel time of the photon being absorbed at a given point. The TT is not a substantial parameter that could be measured directly, since the photon does not have the capability to store or communicate its own TT.

The metrix of a "space" is a parameter that was fully analyzed for central velocity fields with only one emitter. The consolidated effect of photons arriving from diverse directions cannot be found by an algebraic addition; each particular photon communicates a relative velocity on a particular direction. The velocity means a change of position, which in turn means that the particle will reside in other point on the "velocity field", which in turn means that it will move at other velocity and to another direction. When this process involves the sequential arrival of photons from various sources, every photon will apply separately a force to the receiver, which increases or decreases its distance to that particular emitter; therefore, the sequence of solicitations will modify the final result, and the sequence depends on the quantity of photons arriving from each emitter, which is proportional to the brightness of the emitter. Consequently, the time integral defining final position will not only be a directional function but has also to account for the relative brightness of emitters. This example illustrates how a composite "space" involving many sub-spaces is not the simple vector addition of velocities assigned to each sub-space point where the receiver resides; moreover, when the emitters move with respect to each other, then the problem increases in complexity.

When analyzing the movement of self propelled objects in an intergalactic space, one must be very careful in applying the concepts of "space" and "velocity." Under this circumstance, there is not a clearly defined "velocity field", so integrating it in order to find a "space", does not have any meaning.

## The aging effect in photons

The following proposal will be referred as "aging effect"; it states that *"in any Hyle structure the TT unit grows in relation to AT unit (the second) as time passes"*; let's analyze the phenomenon.

The "law of symmetry" ($t = \tau$) institutes the "Hooke's Law" as the fundamental property of Hyle structures; which in turn, is the origin for all basic natural parameters as: force; work, energy, gravity, etc... This law rules that the internal elasticity corrects automatically any deviation from the equivalence of particle period and length; when applied to the behavior of a compressed photon ($\Delta T_o$ / $T_o > 0$), originates gravity as the fundamental property of space, which is generated by the asymptotic elasticity of photons recover nominal length. The same law rules the mechanics of absorption and emission, when the particle ring perimeter is larger than the period that decreased because of the absorption.

Surprisingly[142], the nature dictates that the photons length (or any Hyle structure), expands permanently when detected after it travelled for a very long time; the problem is that the particle does not offer an elastic counter reaction in order to recover nominal length as established by the "law of symmetry." This behavior apparently contradicts the law of symmetry ruling that the ratio ($T_o$ / $T_o$*) should always be one, but as explained, the photon length grows permanently to values larger than one.

Attributing this unbalance to a change of the absolute time reference established by period '$T_o$*', is unacceptable; so, the only alternative left is that the observed photon length, as measured by the receiver, is larger than the one the particle internally reads, so no elastic forces will be generated, and the internal metrix will remain as one. This explanation implies that the INTERNAL TRAVEL-TIME **UNIT [TT sec]** grows as TT increases, so the ratio of length and period remains being one ($T_o$ / $T_o$* = 1). The growth of TT unit [$\tau^o$] relative to AT unit [$t^o$], then is:

$$\tau\,[\tau^o] = t\,[t^o]\,(1 + \sigma\,\tau) \qquad (17\text{-}13)$$

And: $\quad T_o / T_o^* = \{\,To\,[(1 + \sigma\,\tau)]\,/\,(1 + \sigma\,\tau)\}\,/\,T_o^* = 1 \qquad (17\text{-}14)$

Under these conditions the internal metrix remains being one because the internal TT is measured with an expanding scale which never perceives the increase in length perceived at reception[143]. Assuming that this behavior is also applicable to material particles, we must consider that the ring is a closed loop structure which will correct the very small asymmetry generated by this rule, as soon it absorbs and emits a slightly more energetic photon, so correcting any asymmetry;

---

[142] A profound analysis defines that it is not so surprising because the distance of galaxies could not be defined by symmetric photons.

[143] When a particle absorbs an aged photon, uses the absolute time as reference to measure the length of the arriving photon; while the photon is not aware of this change.

consequently, this phenomenon will only affect to "open structures"[144], as a photon.

Every free particle in the universe is an independent entity, so the rules affecting its properties may not affect others; therefore, the aging principle affects to every particle individually, so the free photon expands its own scale as the rest of the particles in the universe may not be affected in the same manner. This important fact may explain the origin for the observed expansion of the universe; material particles are not affected by the "aging effect"; moreover, the atoms in large material objects will not change metrix because of their intensive signal exchange. The different behavior of the open and closed structures gives the potential for space for expanding.

It is interesting note that the energy '$\varepsilon$' is a parameter derived from the force 'F' applied to change the position of a material object a distance 'd' ($\varepsilon$ = F d). in the case of a photon, it defines 'd' and its elastic structure is the origin of force; therefore, when its length '$T_o$' is the nominal length, then the distance 'd' is zero (d = 0), so a symmetric photon does not generate any force 'F'; consequently, the energy consumed by the phenomenon of space expansion is zero ($\varepsilon$ = 0). Under classical semantics we may say that space curvature becomes concave with no energy expense[145].

One phenomenon which deserves to be commented is the one that refers to the observed recession of distant galaxies going away at speeds higher than "light speed" 'c'; classical physics cannot explain how we can detect signals generated by objects getting away faster than light, because the TR denies the possibility of material objects

---

[144] I am not absolutely satisfied with proposed solution, but I cannot find another explanation. Under this rule, nature dictates that the travel-time clock adds a delay of $2 \cdot 10^{-18}$ seconds per every second of elapsed absolute time.

[145] A photon travelling through space does not consume, neither produces energy; it communicates a "potential" energy by producing elastic forces in the receiver. The "internal photon energy" is only related to its angular velocity, not to the length..

travelling faster than 'c'. On the other side, the TH considers that photons travel away from the emitter at the rate of one length per revolution, so setting the metrix (and "light speed") in that location of its sub-space; consequently a photon communicating a negative speed larger than 'c' will travel also a longer distance per revolution. In conventional terms old photons travel faster than light speed in a flat space, they travel exactly at the same speed that they communicate to the receiver.

## Gravity inversion distance

The distance '$\tau_\xi$' at which gravity inverts its sign will be found, when the attracting sub-space G speed ($v_g = c\, A_g\, /\, \tau$) is equivalent to the space dilatation speed ($v_h = c\, \sigma\, \tau$):

$$\sigma\, \tau_\xi^{\,2} = A_g \tag{17-15}$$

$$\tau_\xi = (A_g\, /\, \sigma)^{1/2} \tag{17-16}$$

As ($A_g = G\, M/c^3$),  $\quad \tau_\xi = (G\, M/\, \sigma\, c^3)^{1/2} \tag{17-17}$

$$r_\xi = (G\, M/\, \sigma\, c)^{1/2} \tag{17-18}$$

From equation (294):  $\quad r_\xi = (G\, M/\, Hu)^{1/2} \tag{17-19}$

The gravity inversion distance depends on many other factors [146] such as emitter mass, its shape, and the geometric distribution of other emitters in the vicinity. This phenomenon may explain how the size of galaxies is defined by nature and why the universe is not a single gigantic galaxy; the rotational mechanics of galactic arms will be affected by this gravity inversion. A detailed study may find some relation with observed anomalies currently suggested as caused by the existence of a "dark matter."

---

[146] A coarse calculation results in a distance ($r_a > 3$ Kpsc); which is similar to Milky Way diameter.

## The speed limit

The TH shows a much more comprehensible portrait of a galactic universe. Previously, under the paradigms of the Theory of Relativity it would not be acceptable admitting the possibility of traveling faster than light; so human kind would be sentenced to be caged in an arm of the Milky Way; the rest of the universe was senselessly instituted as an inaccessible object, being there just to be observed. It is important note that space metrix depends on the sub-space characteristics, as does distance; therefore the classical term "speed" should be very carefully applied. The moving object should always move relative to a proper reference, and when this reference object is left behind or is the destination; the magneto-gravitational effect; and photons length should be taken in consideration.

Classical physics demonstrates that an object residing on intergalactic flat space, either static or moving at constant speed relative to an imaginary flat space grid, will never change its relative speed; additionally, there is no physical law restricting to a material object change position in a quasi-flat intergalactic space at the acceleration and speed determined by its reactive propulsion device, pushing it to move in opposite direction to the propellant. The object may be travelling at a velocity several times higher than light speed, and the propulsion will accelerate it as if it would be static; therefore, its dynamics will be fully described by Newtonian mechanics.

Remember that in accordance to the TH, a material object is absolutely isolated when it does not receive any signal, and the emitted signals generate its own sub-space which is always "carried" with it, because the emitted signals will always get away from it, as if it would not be moving. Properly explained, moving away faster than 'c' from a reference object means that the communication delay from a photon emitted by the reference source is increasing faster than absolute-time ($\Delta\tau > \Delta t$).

Even the conventional definition of speed should be adjusted to this situation; for instance: the propellant material left behind could be used as reference because the ship detects its photons while its relative velocity remains lower than 'c'.

A spaceship travelling away from the sun at a speed 'v' higher than 'c', will leave behind sun photons and will intercept the ones which preceded it, "from behind", so it will perceive initially a "very low frequency radiation" in front of it, as if the sun appears being ahead; afterwards, the frequency of absorbed photons will increase as it accelerates. When 'v' decreases, the sun will reappear behind as a small point. The only permanent reference will be the signals emitted by the destination object, and the velocity concept only makes sense when referred to the propellant.

Initially when (v = c), the photons emitted by the destination object "A" which recedes at a speed larger than 'c' ($v_d > c$), will indicate that it is traveling away at a speed ($v_d - v$); '$v_d$' will decrease as the spaceship approaches to it; when the perceived velocity is the same than 'v', the spaceship will enter the gravitational subspace generated by "A".

To compute the travel time of the spaceship one should consider other effects (*)[147] as:

---

[147] The classical concept of space, distances in a three dimensional environment, is a mental abstraction; a more real representation will be to imagine a velocity field (gravitational sub-space) as water running towards a drain in the center of a flat container whose base is calibrated by a square matrix of unit distances representing the geometric abstraction of an ideal flat space. The expanded sub-space will be the opposite; it will be represented as a water source in the center of a spherical surface, so the water runs faster, when it is farther away. Any material object resting on it will be like a boat whose relative velocity is the one set by the water current that point. When the boat rests on a "expanded space", and it is propelled to approach the water source at a velocity 'v', it means that it must travel upstream faster than the water flows in opposite direction. When the boat enters to the gravitational sub-space, it will accelerate because the water runs faster as it approaches to the drain (gravitational attraction plus magneto-gravitational effect).

a) The velocities from the subspace of destination mass, add to the relative "speed" of sub-space at that point; initially the destination point recedes while the spaceship travels in its "expanded sub-space" and later approaches when it enters to the gravitational sub-space.

b) The "magneto-gravitational" effect is observed when the ship approaches to the target following a radial path in its sub-space; the sub-space movement increases the relative velocity, therefore it accelerates faster than gravity.

c) The acceleration given by the self propulsion device.

## Black holes

A black hole is a mathematical entity generated when exploring the limits of Theory of Relativity equations [148] that consider space as substantial natural elastic entity affected by the action of gravity; on the contrary, in accordance to the TH both parameters, space and gravity; are contingent entelechies found by respectively integrating or deriving the "velocity field."

Therefore, space is the mental abstraction of the neighborhood of a massive emitter with photons passing permanently through it, establishing naturally a velocity field, from which is developed a secondary abstract concept that is a position field (space), and other abstract concept of acceleration field (gravity).

The theoretical proposal of an entity having a large mass which produces a strong gravitational field that attracts everything, even photons, is not possible under the TH perspective. Gravity and space are two properties describing the effect of myriads of photons emitted

---

[148] Any mathematical result exploring the limits of a natural phenomenon is prone to apply equations outside of their validity limits; they do not reflect nature reality under those extreme values.

by the massive body; there cannot be gravity when not photons are emitted. Large gravitational fields can be generated by massive dark objects emitting a few photons; enough for generating a stable and permanent sub-space.

## Neutron stars

The concentration of similar particles of the same size could be theoretically possible, but they will not generate a gravitational field because of the following facts:

- Neutrons are never completely neutral type T particles; their asymmetry involves that they always contain some R and V components; so it is highly speculative assume that they will get together easily.
- Neutrons normally do not emit photons; when they break, they split in two opposite sign particles: a proton and an electron.
- A pure neutron star cannot produce gravity; because, similarly to an ideal "black hole" it would be an entity that does not emit photons, so it does not communicate with the material universe; it becomes an undetectable entity that cannot influence, nor communicate its existence to its vicinity, unless all its neutrons bond together like a single atomic nucleus; only in this speculative case they could emit highly compressed photons.

## Gravitational disturbances

The TR assumes that space is a natural substantial entity with elastic properties whose local dimensions are affected by the presence of matter. On the contrary, the TH considers space an abstract object that eventually may cease existing when it does not contain low energy photons passing through it; these photons must be absorbed by the material objects close to the emitter.

A sudden gravity variation may be produced by the partial or total collapse of an emitter; That is, when their rings are converted to photons. A massive emitter destroyed by powerful gamma rays, will not emit low energy photons anymore; so its sub-space will suddenly cease existing; consequently, any receiver residing on it will cease absorbing those photons, so ceasing abruptly the action of gravity forces to which it was subject; this will produce elastic changes in receiver structure; this effect is directly related to the elastic deformation of the receiving object, not by "space waves."

## Creation of the universe

The sequence of events in the beginning of time from the perspective of the TH differs from the currently most accepted proposal known as the 'Big Bang', which is strongly based in parameters as: high temperatures, large densities, gravity, space, etc. In accordance to the TH, all these concepts cannot be properly applied when there is not matter, not space, not electric forces, and there is not gravity.

The TH proposes a probable scenario; a comprehensive, deeper investigation could confirm it:

## 1) LIGHT

In the beginning there was no space, nor time, only a potential source of Hyle, which is capable to emit photons in all directions.

Suddenly, a point source emits high energy [149] asymmetric photons, vibrating probably at the frequency of neutrons. The high energy photons spread out, each one following a straight "radial trajectory." The term "direction" is difficult being properly applied to this kind of environment because there is not a physical space yet, there are only possible dimensions; consequently one may assume that some of the

---

[149] Energy is a term whose final meaning is the capacity to move material entities, but with no material entities, nor space, there is no distance, nor directions.

photons follow curved patterns as being straight lines in a curved space; others will curl up in a "circular" path; forming neutrons. The duration of this process defines the initial "size" of the universe; because the photons following a radial trajectory will generate distant neutrons at larger travel-times from the source.

## 2) INFLATION AND CHAOS

Communication among particles is not possible because "space" does not exist in this inflating "volume" that increases the travel-time distance between neutrons colliding with high energy photons which are all over it.

The neutron, a type T structure, is the key which generates all other particles. The phenomenon of neutron decay explains how a neutron splits in two "charged" stable particles; when an x-ray photon is released by the neutron follows a highly curved path, it produces an electron, and leave a proton. The described mechanics explains the genesis of atoms and matter emerging from neutrons splitting in identical quantities of protons, and electrons.

## 3) GRAVITY AND ELECTRICITY

While the chaos continues, some of the newly formed protons collide with neutrons, eventually getting together shape an atomic nucleus. In this stage, strong proton sub-spaces E dominate the "volume" that captures the emerging electrons, constituting atoms of Hydrogen, Deuterium, Tritium and Helium.

When the previously described chaotic stage decreases its destructive activity, a continuous signal interchange starts with the presence of non destructive signals that slowly prevail, thereby establishing diverse sub-spaces G which accumulate matter.

## 4) **STARS AND GALAXIES**

The birth of Stars; planets;, and galaxies. The creation of individual galaxies, and not a single giant galaxy, may be explained by the anti-gravity effect generated by Hubble effect.

## 5) **LIFE**

Matter evolves into complex molecules resulting in living entities adapting themselves to the environment. This process lasts for millions of years, giving birth to virus, bacteria, vegetation, and animal species.

## 6) **SELF CONSCIENCE**

A commanding organ controls animal behavior by processing internal and external information. Then this organ, the brain, evolves and creates a special routine of "self awareness"; giving birth the "free will", a routine of thought that behaves independently from the inflexible commands of physical laws.

In conclusion; *we are material and free creatures that inhabit a deterministic universe born from the powerful light that flashed at the dawn of time.*

This page is intentionally left in blank

**This page is intentionally left in blank**

A



## Comparative review with current Theories

The TH poses a new perspective of nature, but this new point of view does not invalidate the progress made in many aspects by current theories; on the contrary, the concepts expressed by them are enriched by explanations about the origin of some of their assumptions. The TH describes in a more natural way how the material universe behaves, avoiding mathematical speculations about physical reality.

Currently, there is not a theoretical proposal describing entirely nature behavior; most of them are extremely narrow. This applies to: Quantum Mechanics; Electromagnetic Theory, and to the Theory of Relativity; all of them are acceptable within their scope of coverage, but all of them lack the universality in describing nature as a whole.

This is because all of them analyze abstract entities which describe the group behavior from a large quantity of particles; this approach does not always provide the ability to find the essential causes of a phenomenon.

The fundamental assumption of the TH is that the primary cause of any natural phenomena is the interaction among individual particles; through this vision the TH proposes a unifying perspective in theoretical physics. The comparative study here presented, briefly illustrates about the major differences among the TH and main current theories. Under this perspective, the TH establishes a route for the return of the physical sciences to their original course, giving priority to logic comprehension on every particular process, over the mathematical manipulation of contingent abstract entities, as "space", "waves" or "fields" which may only reveal the statistical behavior of groups.

The reader will understand that the critical analysis herein presented, does not cover all aspects of current physics theories; this quest is an assignment for hundreds of researchers, and is well beyond my personal capabilities.

## Historical analysis

Modern physics attitude in exploring nature giving priority to the application of mathematical tools was a consequence of the astonishing development of mathematics in XIX century; since then, visible light was considered a wave included in the electromagnetic spectrum described by Maxwell equations, analyzed with Fourier integrals, obeying Hamiltonians; etc. The constancy of the speed of light gave birth to the Theory of Relativity setting the concept of a four dimensional space-time continuum demanding powerful analytic tools, and the statistical behavior of particles originates the Quantum Mechanics, highly mathematic analytical approach; since then, physics investigation strongly rests on mathematic.

This perspective on nature and the use of mathematics was: adopted, disseminated, taught, and supported in universities, investigation centers and labs. Currently, even matter is pictured as a phenomenon in a multidimensional environment full of waves and fields, which under certain conditions condense into material particles. The priority given to mathematical procedures, over substantial explorations sometimes ends describing mathematical "truths", and not real nature.

## Mathematical tools and physics

The application of mathematical tools in exploring natural phenomena is a procedure commonly established in modern science as a helpful instrument to efficiently investigate the consequences of phenomena generated by large groups of individual entities grouped as: "electromagnetic waves", "space", "light", etc.; this practice drives the advancement of science by exploiting the synergy of thousands of individual intellects[150]; however, this approach is not a substitute for the intelligent imagination assisted by the logic analysis of individual creative minds.

Currently, theoretical presentations such as this one, risk being considered unscientific and purely rhetorical; the prevailing culture finds inconceivable that a supposedly serious proposal does not contain explanations with advanced mathematics, and does not follow established formats.

Of course it's correct use theorems and mathematical operators in finding the consequences of a given equation, the mathematical science have been developed for this purpose and its progress is commendable, but prospecting nature with pure mathematics must always take into account that none of the equations describing physics laws reflect accurately and fully, absolute reality. Any equation is simply an inaccurate and incomplete expression given to an observed phenomenon; these descriptions will always improve in accuracy by progressively including more factors, initially considered as irrelevant. A pure logical process has the advantage of continuously contrasting the results with original postulates, reviewing them immediately, when necessary.

## Experimental data and theoretical predictions

Experimental evidence over a theoretical prediction does not always validate the proposed explanation, because it always includes a preconception dose on the tested paradigms; for instance:

> - The concentration of electrons on the screen in two slit experiment, is the result of the ring shape of electrons; it does not necessarily prove the ondulatory quality of electrons, or any other particle.

---

[150] Modern society need for an efficient research workforce, has resulted in a university education which emphasizes on the acceptance of ideas over creative dissention.

D

- The spatial curvature is the result of photon expansion; not gravity acting on light. Gravitational redshift is the cause, not the effect.

- The probabilistic nature of particle phenomena is a consequence of their geometric characteristics, its ring shape makes them highly sensitive to orientation, is not an indication of indeterminacy.

E

## **<u>The Theory of Relativity</u>**

Both the TH and the TR describe a four-dimensional universe where reside material entities that interact mostly in accordance to the laws of classical physics. The advanced General Theory of Relativity; adds more independent variables as gravity, so it needs describing mathematically the universe with additional dimensions.

The TR postulates the existence of a space-time continuum based on the experimental results showing that "light" always travels at a constant speed equal to 'c' whatever the direction and the relative speed of the transmitter with respect to the observer; obviously, explain the constancy of 'c' which is the ratio of distance and time, both parameters must be adjusted to meet that requirement, whereby instituting the concept of a universe where time mixes with relative positions, this method generates paradoxes and uncertainties.

The TH perceives nature in a different way; the "distance" between two objects which never communicate could be anyone; one centimeter or one hundred kilometers; we will never know it unless we can measure the time it takes communicating them. Their "remoteness" will be given by the time it takes for the information to travel from one to another. The TH establishes that this delay is given by the geometric characteristics of the information carriers, the photons; consequently there is no sense in assigning pre existence to an entity (space) that disappears when there are not signals passing through it, and even more important, whose scale (distance unit) is defined by those signals.

Based on the above conclusion, the TH establishes a single column that sustains the entire universe, the Time; and it places all existent natural entities in a four time dimensions universe.

- The absolute time, a <u>continuously increasing irreversible reference parameter</u> that establishes the foundation for causality, providing also the only "independent, universal reference" to measure the rate of change.

- Three independent travel time dimensions, which specify the travel-time for a signal going from one point to another inside a 3D volume.

These concepts forge the following significant consequences:

- It is not necessary a flat space prior to the birth of the universe[151].

- As the TH does not need considering relativity of (absolute) time, time paradoxes disappear, and the solid concept of a universally causal nature is restored.

- That the signal properties define the spatial unit of length (metrix).

- That 'c' is a conversion factor from TT to metric units. Its units resemble speed, but it is just a conversion factor.

- Consequently, all physical parameters deserve a conceptual review.

- That there is not a speed limit for self-propelled bodies

Deepening on the comparative analysis; the basic difference between the TR and the TH is that the TR considers space as a substantial physical entity, while the TH sees it as a contingent entity that exists only as defined by a mental abstraction when a volume contains

---

[151] This reason has a great philosophical and physical significance, since it states that it is not necessary to locate the position of the universe at the time of birth; it is absurd to locate something when there is no empty space without even one reference. Additionally this philosophical perspective shows the possibility of a virtual universe residing in a creator mind.

material entities emitting signals and their properties are defined by geometric characteristics of the signals passing through it. Most of TR predictions relate to the action of large groups of particles on material entities; obviously the TH results will agree, regardless of the conceptual differences on the origins of the entity under analysis.

The TH explains that the conventional concept of space is the result of transforming of the three TT coordinates in metric units[152], more accessible to human mind. Thus, the distance is determined by applying a conversion factor 'c' to travelled time; obviously 'c' must be a universal constant, reflect the isotropic and homogeneous 3D travel-time dimensions, as observed in nature; one second of the time travelled by a given photon will not always be equal to the travelled time of another, so the "spatial distance" will differ in the same proportion, independently if measured in TT seconds or in meters.

## Space-time continuum

The establishment of a space-time continuum generates the mistaken TR concept that time is relative, by mixing the properties of TT and AT.

Although the TR bases its conclusions on the existence of space as a substantial physical entity, many of its findings coincide with the natural reality[153], while others generate unreal or absurd logical

---

[152] A superficial analysis of this concept may consider that it is an irrelevant issue; a mathematician will say that is a trivial conversion of units; although, actually it involves the redefinition of all physical parameters to accommodate a universe where the particles are the entities which define "distance". Under this perspective, the term "space" becomes a contingent entity, unable to offer a solid reference.

[153] The concept curved space, is the way used by the TR to explain that the relative TT growths as photons expand; the concept of matter shrinking "space" is also a way to explain that the initial photon compression is proportional to emitter mass; the gravitational redshift is explained as the effect, not as the cause. About the paradoxes; the most pernicious effect on physicists is that there is a current cultural tendency to accept as normal the uncertainties, the paradoxes and relativities.

entities; surprisingly science currently ignores some of the absurdities and paradoxes arising from a proposed space-time continuum.

There is an undeniable reality: independent entities occupying different positions, must communicate among each other establish a causal universe; and most important, that this <u>communication cannot be instantaneous, it must have a delay</u>, otherwise all of them will one and will occupy a single point in space.

The communication delay or TT, must be proportional (not necessarily linear) to remoteness; consequently, it does not mean that the position and relative speed will change the properties of space-time, but is simply are a distortion induced by the limited signal speed when observing distant moving objects, because the TT varies when the observed object changes position. The basic statement of the TR is that the relative speed distorts the space-time continuum modifying the "space" or absolute time properties; for the TH is just a distortion induced by the length of the photons.

## The "speed of light"

The conventional way expressing that a photon travels a distance in a given time is done by assigning a "speed" to it. The TH explains that the speed of light is an absurd concept [154] considering that its magnitude ($c = s / t$) uses the parameter 's', whose magnitude is defined by the same "light", so this definition mixes causes with effects; even worse, it uses the effect in order to define the cause.

---

[154] Velocity is conventionally defined as the ratio of the distance traveled by an object during the time interval perceived by the observer. Obviously this ratio is subject to the inaccuracy induced by the variation in travel time of the signal as the object changes positions, and the unknown magnitude of the "distance" travelled in that point of space. The "distance" "A-B" is defined by the TT of signals travelling from "A" to "B"; therefore, the "speed of signal" it is being defined by using a parameter than the same relationship sets; "distance" is defined by the TT of the signal and the "speed of light" 'c' is found by that distance with reference to the perceived time; consequently, the concept of "speed of light" is an absurd; additionally, an additional uncertainty is generated because the speed definition does not specify which of the dimensions of time (absolute time or travel time) should be applied.

The alleged bond between space and time when defining the concept of the continuum, is the primary cause for the TR mixing the properties of absolute-time 't' (independent and irreversible) with the ones of travel time 'τ' (reversible, and variable with spatial position); so setting grounds for the TR presuming "the relativity of time."

Whereas all communication is done through information carriers (single photons), the perception of change in relative position of the observed object, whether it concerns to velocity or acceleration, depend solely on the information the signal communicates to the observer. When the observed object approaches, the perceived signal frequency is increased by the well known Doppler's effect; when the frequency increases as the emitter approaches, it indicates that the object accelerates. This fact shows that acceleration is independent of the source because: the distance unit, velocity and acceleration, is set by the same signal. For these reasons, it does not need establishing a "principle of equivalence" because gravity is acceleration set by the signal that communicates a progressive increase in relative speed.

## The gravitational red-shift

The general TR considers that gravity is a natural phenomenon affecting all natural entities, including photons and electromagnetic waves; in other words it posits that gravity attracts light, and it shows with the experiment showing the "gravitational red-shift" of monochromatic light depending on its distance to the gravitational mass. In conventional terms it declares that the loss of internal energy of the photon is converted into potential energy.

Explaining the mechanism described under that assumption, it seems that a photon moves away at the speed of light with a gravitational field which also spreads at light speed, and interacts instantaneously (?) with the photon; so increasing its wave length as the gravitational field gets weaker. This illustration contrasts with the simple assumption of the TH that the gravitational photon carries information. The TH proposal presents a much a simpler and direct explanation on the red-shift by considering that this is the cause of the gravitational phenomenon; the photon is the carrier of information that informs the receiver that the emitter is approaching continuously increasing its speed as it gets closer. This explanation demonstrates how, depending on your point of view, the same experiment can be interpreted as evidence that demonstrates the validity of two conflicting hypotheses. The difference is that the TR does not explain the causes of the gravitational phenomenon or the paradoxes arising from it; while the TH uses this concept deduce Newton's equations.

### Gravity

For the general TR, gravity is the effect of the curvature of space-time continuum. For the TH is the relative velocity communicated by the absorption of a compressed photon; or in conventional language, is the effect that induces the velocity field around a material object. Both definitions declare the same fact, but each one uses its own semantics.

For the TH the "relative velocity field" generated by an emitter modifies the "metrix", so defines a gravitational acceleration on every space point. The TR explains the same by saying that the "space-time continuum" gets curved. The profound difference is that the TH demonstrates that "space" and "gravity" are the two faces of the same contingent entity generated by large groups of photons; space is the integral of the "velocity field" and gravity its derivative.

Considering that "space" is a physical stable entity when there are photons passing through it permanently, the geometry of this environment will be defined by the preponderant direction of the velocity vectors in this field, so the straight line for a tangential trajectory will be a circle; this is the case of the observed phenomenon of a gravitational lens; the photons from a distant star travel a circular path when they approach a massive emitter.

## Mass energy equivalence

The TH proposes that the only raw material of the universe is Hyle, which takes different forms for constituting matter and signals; the TR statement that mass and energy are forms of the same natural entity is absolutely coincident with the TH proposal.

L

This page is intentionally left in blank

## The Electromagnetic Theory

The ELM (Electromagnetic Theory) proposes that electromagnetic oscillation is a substantial natural entity which transports energy travelling at light speed. These waves cover the whole spectrum, ranging from gamma rays to low frequency radio waves.

The TH considers that there are two differentiated categories of oscillations; one referred to the substantial property of Hyle to oscillate, only related to the energy contained by a particle (Einstein-Planck relation), and the space-time phenomena involving large groups of particles; which is the subject analyzed by the ELM.

The ELM, like the TR, is firmly rooted on abstract entities as "space", "current", "flux", "fields" and others, referred to phenomena compromising large groups of particles travelling and interacting among each other; on the contrary, the TH is centered in the analysis of individual interaction between two particles.

The ELM wave equations are closely related to the classical concept of 'c', the "light speed", generating similar conceptual imprecision than that of the TR. On the other hand, the TH considers that 'c' is a conversion factor to transform travel-time magnitudes to artificially instituted metric units; and most important, the photons carrying the electromagnetic waves define the "distance." The fact, established by the TH is that the universe resides in individual "velocity fields"; not in a metric "space."

The ELM Theory has its origins on the equations ruling the behavior of groups of moving electric charges by integrating, in various forms, the two basic electric theory equations: the Coulomb and Biot-Savart laws; resulting in "fields", "electric and magnetic fluxes" which are orderly shown by the well known Maxwell equations.

The TH considers important to mention that sensors normally contain large groups of receivers; therefore, the detected frequency is the average of the frequency from a continuous train of photons with similar frequency. There is another kind of detection; when a large receiving object (antenna) decodes a modulated signal; recovering the sequence and the spatial distribution communicated by a large group of photons orderly emitted, and sequentially received. In both cases every photon in the group travels on the straight path defined by the sub-space, which are individually absorbed, reflected or refracted.

The following analysis shows that the mathematical description of the behavior of a particle or group of them moving in a central sub-space is a mathematical abstraction with geometric consequences, which could be applied to any natural phenomenon of a moving entity attracted by a central force; it could be electric, gravitational or even a group of corks floating close to a drain.

## The electromagnetic phenomenon

Currently, the scientific community accepts that electromagnetic forces are substantial natural forces, given by the strong experimental confirmation of the phenomena described by the four Maxwell equations whose development was based on the basic laws of electricity and magnetism; the ELM currently is a dominant branch of physics.

The irrefutable proof of the importance given to the Electromagnetic Theory by modern science is revealed by the universal institutionalization of the spectrum of electromagnetic radiation, widely considered as an explicit demonstration that exists a substantial natural ingredient shown by electromagnetic "waves" propagating through "space" in the whole spectral range which covers the all known oscillations; from "gamma rays", visible light, and low frequency radio waves. As a direct consequence of this situation and considering the universality of mathematical truth expressed by Maxwell's equations supported by extensive experimental evidence, is not difficult to understand why magnetism and ELM have become widely accepted as a basic natural substance.

The TH shows an extremely important fact; that the electric attraction/repulsion applied to a charge results from its interaction with distant charged particles is a local phenomenon taking place when the receiver absorbs the polarized photons (Pp), changing position by accelerating towards the emitter or moving away from it. This electric force is proportional to the gravitational pull, since both are inversely proportional to the square of the distance ($F = k / r^2$); this is like saying that there is a factor "$\eta$" ($\eta = k_e / G$) that multiplies the action exerted by gravity among unitary masses, in relation to the action of the force among unitary charges at the same distance 'r'. Consequently both; gravitational and electric forces are similar in substance, but different in magnitude, so whatever other property, expression or behavior applied to one, must be common to both; consequently the Maxwell equations must also describe "gravity waves." This assertion is proved by the TH when demonstrates that "B" is a mathematical parameter describing the effects on a receiver moving on a "velocity field"; there is an additional force accelerating it as expressed by the laws of Biot-Savart and Lorentz.

Therefore, the TH demonstrates that the presumed "magnetic field" 'B' is simply an intermediate variable [155] used for facilitating the mathematical analysis of complex varying central velocity fields;

---

[155] The TH demonstrates that the "magnetic field intensity" 'B', originates when a charged particle travels in "velocity field" which induces a velocity change (acceleration) whenever the charge changes its distance to the emitter ($a = dv / dt = dr / dt \, dv / dr = v \, dv / dr$). When this speed is not purely radial, the term ($dv / dr$) corresponds to the quotient of two vectors, which results in another vector 'X' perpendicular to both ($X = v / r = v \times r / r^2$), whereby acceleration can be written as ($a = v \times X = v \times v \times r / r^2$) which is the equation of Biot-Savart (also known as the Lorentz force $F = Q \, V \times B$), showing that 'X' (or 'B') is only the name assigned to a variable which results from the ratio of two vectors; the ELM theory assigns the name of magnetic 'B' field because it is associated with the force that takes place between two magnets or solenoids. This analysis proves that magnetism is not a physical entity, it is just represents an intermediate mathematical variable generated by a vector ratio.

electric, gravitational or of some other kind. The magnetic field "B" it is not an independent physical entity, neither a natural substance[156]; it is simply an abstract, contingent mathematical entity; these conclusions also apply to gravity fields.

In conclusion, the basic electromagnetic equations are not a particular to electric phenomena, but they are a general description on the behavior of individual particles, or groups of them, moving in central velocity fields. The mistaken assumption which enfolds these phenomena, assuming that it is only applicable to electricity and magnetism, and not to gravitation, is understandable considering the easiness of verifying experimentally some results in a small laboratory using groups of charged particles traveling through a conductor, versus the difficulty of executing gravitational experiments with self-propelled vehicles moving through interplanetary space.

A second important issue is, clear away the confusion of denominating "light" to a group of photons travelling away from the emitter, where each one oscillates at a particular frequency. Under the ELM perspective, light is an "oscillatory substance" detected as "electromagnetic waves"[157].

### Maxwell equations

Maxwell's equations, are actually the algebraic formulation of general mathematical theorems applicable to any natural phenomenon related to central "force fields"; the Maxwell's equations describe the magnitude variation of abstract parameters (E and B) in relation to spatial position and time; always assuming that the "space" is a

---

[156] It is extremely important to observe the fact that a natural sub-space is not a metric space, but a "velocity field"; consequently the integrals and derivatives, which are always related to positions, implicitly affect relative speed.

[157] It is important to always remember that the concept of: light, field, space, electromagnetic wave, etc.; are names given to abstract entities containing a large quantity of particles.

substantial natural entity, used as a firm reference, and not the entity whose properties are defined by the qualities of the photons constituting it.

From a mathematical point of view, Maxwell's equations are simply descriptions of the space-time geometric effect, from a group of emitters on the receptors located in its vicinity. James Maxwell, Oliver Heaviside, Heinrich Hertz, Ampere, and Faraday, developed this theory by finding experimentally the equations governing the interaction between moving charges; but all of them are a consequence of the equations set by Coulomb, Biot-Savart, and Lorentz. The result was instituting magnetism, as a natural parameter, related to electricity; when in fact magnetism is a mathematical variable conveniently used for analyzing the forces on moving charges. The mathematical analysis of a central velocity field is not only related to electricity, but also to gravity, or to a floating object navigating on a fluid close to a drain.

Obviously it would be improper say that electromagnetic waves do not exist, because its mathematical description is accurate and true, but it is wrong assume that these waves are a substantial physical entity. It is indisputable the brilliance of Maxwell's equations and their implications, which surprised even to Maxwell; but, these equations can be applied to any phenomenon related to a set of emitters sending coordinated periodic waves, as in the case of the radio waves; but they could also be applied to periodic gravitational waves emitted by controlling appropriately the positions of several massive bodies.

Historically, these equations gave birth to an era of "mathematical physics" which persists nowadays, rooting research on mathematical operators, dismissing researcher's confidence in their own logical reasoning and instituting an almost dogmatic attitude of blind faith on the results given by mathematical manipulations; many times offering grounds for assumimg that these results represent a natural substantial phenomenon. The ELM theory is a clear example about the exaggerated preponderance of mathematics in modern physics;

currently is acceptable considering electromagnetic waves as a fundamental natural ingredient.

On the contrary, the TH shows that electrons and protons are the entities that emit photons, they do NOT emit electromagnetic waves; and that those photons are individual entities, whose oscillating substance is not electromagnetic; neither can it be described by Maxwell equations.

The TH also demonstrates that the term "B" is not a natural substance. Therefore, the laws governing distant interaction between charges are clearly described by the equations of Lorentz and Biot-Savart[158], which detail the forces generated by:

a) The attraction/repulsion generated by direct action of the "velocity field" surrounding the emitter (Coulomb), and

b) The additional "magnetic" force applied to the receiver when it it changes positions in the "velocity field." All the ELM equations; Ampere, Faraday, Lenz, and Maxwell, are all a secondary consequence about the fact that "electric fields" and "gravity fields" are "central

---

[158] The TH shows that the so-called Lorentz force is the result of the position change of a self-propelled receiver travelling on a straight path at a speed 'v' in an electric field 'E' (velocity field) because it changes the velocity at which it is subjected by the field when the receiver path changes its radial distance to the emitter; to maintain a straight path (as perceived by an external observer) the receiver should offset the higher speed of the field near the emitter with an opposing force proportional to the approaching acceleration. The foregoing reflects the wording of Biot-Savart law ($\mathbf{F} = [(\mu_o\, q^2)/(4\,\pi)][\mathbf{v} \times (\mathbf{v} \times \mathbf{r})\,/r^2]$), which relates the magnitude of a "magnetic" force applied to a receiver when it cuts the potential surfaces surrounding a central transmitter at a speed 'v' and a distance 'r', this relationship was easily confirmed by old experiments using rigid metal conductors which forced the straight trajectory of the charges. The term $[(\mathbf{v} \times \mathbf{r})\,/r^2]$ represents the so called "magnetic field" 'B'. The induction phenomenon, described by Biot-Savart's law represents also another face of the same phenomenon since a moving emitter, with equivalent relative velocity, produces also its own electric field that changes magnitude and direction of the velocity field at the point where a stationary receiver resides. Classical physics denominates it as "induction", which also has its origin in changing relative position of charges embedded in a velocity field. Exactly the same behavior may be observed when a moving mass goes through a gravity field; therefore the "magnetic induction" is not solely related to electric fields.

fields." All the laws of electricity and magnetism are valid; but the contribution of the TH reveals their origin.

The following lines show the pure mathematical essence of the Maxwell equations:

I) The first two Maxwell equations $(\nabla . E = 4\pi\rho)$ and $(\nabla . B = 0)$ are the law of divergence, known also as Gauss's law, establishing the existence of a central emitter. The divergence is a general mathematical law establishing that the net "flow" (of a group of entities) coming out from a closed surface enveloping a volume, is equal to the net emission capacity hosted inside that volume[159]. For an electric field it is proportional to the total charge contained; in the gravitational case is proportional to the contained mass. In the case of the magnetic field, the equations shows that there is not a natural source of 'B', so confirming what the TH proposes; this parameter is only a mathematical intermediate variable facilitating the calculation of the dynamics among moving charges; consequently its divergence is zero. This type of equation can be applied to emitters capable of generating central forces such as gravity, a swirl, etc. The general equation is $(\nabla . A = b)$, where 'A' is a vector and 'b' is a scalar; it may be applied to various fields as:

    d. In electric fields $(A = E)$ and $(b = \rho / \varepsilon)$

    e. For the magnetic field $(A = B)$ and $(b = 0)$. It shows that there are no individual natural emitters of "magnetic fields" (mono-poles) because 'B' is not a physical entity, it is only a mathematical variable

---

[159] This mathematical truth includes a not so explicit connotation, that natural radiation could be unidirectional; detail to be accounted for in the case of "wave emission". It is also important to mention another detail, that the individual emitters change from one to another, they are never the same because the "charges" in a material object relate to the existence of unbalanced atoms hosting free charges; this unbalance passes from atom to atom, and from a free charge to a new one, because the particle which emits a signal is the one that just absorbed a photon exciting it; so, this is always an statistical process.

    simplifying the geometry of the dynamic phenomenon among moving groups of charges.

    f.    For the gravitational case (**F** = **Vg**) where '**Vg**' is the gravity potential; and (b = mass density)

II) The third Maxwell equation ($\nabla x E = \partial B/\partial t$) is the differential version of Faraday's equation ($\oint E \cdot dl = -ddtB \cdot dA$) modified by applying the Stokes Theorem in a three dimensional space where there is a continuous vector field 'E' and the dynamically related 'B' vector field. But the Faraday's law, on its turn, is just an alternative mathematical expression the Biot-Savart law (also known as Lorentz law) which describes the relation of between the geometric properties of the electric "field", and the induced change of the vector "B" when the receiver moves in the radial direction. The TH demonstrates that Biot-Savart law is also a universal theorem describing the effects on objects residing in "velocity fields"; consequently the third Maxwell equation describes in a simple mathematical form that any directional change of the direction of the vector field 'E' when the receiver location will change the magnitude of the intermediate variable 'B' proportionally to the receiver radial velocity; This assertion applies also to: gravitational or hydraulic phenomenon.

III)    The fourth Maxwell equation that in its differential form is ($\nabla x B = 1/c(4\pi J + \delta E/\delta t)$) illustrates also <u>a mathematical truth, not particularly related to a specific physical entity as the "electric field."</u> This equation shows the behavior of a group of charges when moving on a varying electric field 'E' related to the parameters: "B" and "J". The same relation could be applied to masses moving in a varying gravity field, or to hydraulic phenomena where the volume of

> drained liquid changes with time; they also describe a general behavior, applied in this case to a moving receiver in a varying velocity field.

Therefore, we may conclude that the four Maxwell equations are expressions of mathematical theorems relating forces on objects residing on a sub-space (velocity field); these relations may describe: a mechanic, electrical or gravitational phenomenon.

## Electromagnetic waves

Further manipulations on these equations, applied to oscillating sources, result in the well known pair of "electromagnetic wave equations", which describe the space time relation of electric and magnetic fields:

$$\frac{\partial^2 \mathbf{E}}{\partial t^2} - c_0^2 \cdot \nabla^2 \mathbf{E} = 0$$

$$\frac{\partial^2 \mathbf{B}}{\partial t^2} - c_0^2 \cdot \nabla^2 \mathbf{B} = 0$$

The symmetry of the above two equations evidences that there is a spatial dependence of the intermediate variable 'B'[160], with the vector distribution of the primary variable 'E'.

These equations give the magnitude and direction of the abstract vector "**E**" defined as the "electric force field", and the magnitude of the intermediate variable "**B**" at a point is a space, when the source varies with time the magnitude of its emission. Obviously they are valid mathematic relations; but they do not explain, neither imply, about the origins of the natural phenomenon. So it is improper assign natural substantial qualities to the magnitudes to these abstract parameters which only offer the average values about the local effect

---

[160] The electromagnetic wave equations are general mathematical expressions defining the direction and magnitude of "E" and "B" in a point of an electric sub-space generated by a point source that changes its magnitude with time, indicating also that the orthogonal vector 'E' and 'B' are synchronic.

of groups of Pp on groups of receivers. The results given by these equations are particularly useful work on modulated emissions of Pp.

The above analysis confirms that there is not any valid reason assigning electromagnetic wave properties to individual photons; as shown by the TH, their frequency is an innate property of its oscillating substance; the Hyle.

## ELM spectrum

Maxwell "wave equations" describe with precision the behavior of electromagnetic oscillations travelling through space, giving rise to assume that natural entities communicate using electromagnetic oscillating signals, concluding that "light" is also an electromagnetic wave, and all oscillatory radiation travelling in vacuum is ELM; furthermore, some theoretical propositions even go farther, describe the photons as electromagnetic wave entities.

Moreover, it was experimentally confirmed that the Maxwell equations describe the distant effects of informative signals by arranging a sequential spatial distribution of charges in radio emission[161]; but this success does not mean that the emitted signals, initiated by electromagnetic manipulation at the emission point have to be "electromagnetic"; the TH shows that the real description of the phenomenon is that some emitter points generate a time varying "sub-space" by sending a controlled quantity of polarized photons (Pp) which travel in groups forming a "composite-space" with specific properties (magnitude and direction of the electric attraction) in every one of its positions.

---

[161] Radio waves detection is provided by a conductor material producing free electrons at the location where a photon is absorbed. These free electrons produce an oscillating current that changes magnitude at the same rate than the sequential arrival of photons. The frequency of the photon provides the necessary energy to liberate the electrons, and does not have any dependence with the frequency of the radio wave.

The space-time properties of this "composite-space" or "integrated electric field" are precisely described by the Maxwell equations; but it does not mean that the signals emitted have to be electromagnetic in essence; all it means is that the equations describe precisely the magnitudes and geometric properties of the electric pull on a receiving particle; in order to achieve this target, the ELM solves appropriately the problem, by using  the variables 'E' and 'B' describing abstract "fields" referred to the also abstract concept of "space." This does not mean that radio waves are a natural substance, neither light has to be an electromagnetic wave[162]; in both cases the phenomenon is related to the geometrical behavior of a group of travelling photons.

Consequently, the electromagnetic spectrum describes two overlapping segments:

a) The high frequency segment shows the increasing energy sequence of individual photons; possibly it is not strictly continuous, and

b) The low frequency segment, including only the frequencies of modulated photon groups; starting with the X-rays, reaching to the long radio waves.

---

[162] For instance, the emission of white light is a blend of photons of different wavelengths emitted by many atoms with different centrical levels; the same is true with x-rays or gamma rays.

X

**This page is intentionally left in blank**

## **The Quantum Physics**

Both the TH and Quantum Mechanics (QM) posit that the particles are oscillatory individual objects.

One of the main QM conjectures, suggest vaguely that the particles are oscillating entities which interact as individual entities because they are "quantized"; while the TH proposes that they are separate individual entities formed by an oscillatory substance, the Hyle. This proposal is supported by a rational assumption that communication between material entities necessarily implies the existence of individual finite information carriers: they must be finite, for emitters not collapsing with a continuous emission, and they must be oscillating entities residing in a four temporal dimensions environment, be able to carry information; it would not be possible communicating sufficient information to the receiver by entities whose sole properties are: their amplitude or height, and their duration or length. Both the TH and the QM describe particles as a "wave function" $\Psi$ (t, x), but from different perspectives, as explained below.

The TH describes a physical object as a specific vector function H (t; $\tau$), which synthesizes the required properties of particles communicating information from an emitter to a receiver[163], while the QM proposes a general "mathematical entity" known as "the wave function" $\Psi$ (t, x), which may be subject to mathematical manipulations to match its characteristics with experimental observations; the difference is that

---

[163] The TH assumes that the communication is the cornerstone to build a causally universe; consequently the carrier structural properties must be defined by the emitter qualities and its absorption must be able to modify the properties of the receiver, always ruled by natural laws, so establishing a causal relationship between emitter and receiver.

the TH describes a physical object, while the QM describes a general mathematical entity that manifests as matter or wave, dogmatically accepting this unexplained duality. Under these foundations, the QM applies profusely mathematical tools for investigating particle behavior, finding many valid solutions which are coherent with reality; but also finds some absurd results that contradict the basic logic postulates.

In short; based on the fundamental assumption that the universe is an entity residing in a multidimensional time environment where all the objects are made with a single substance, the TH proposes a mathematical relationship describing the physical characteristics of every particle as a separate entity and explains how the proposed model reveals the phenomena of gravity, atomic geometry, electromagnetism, etc. On the contrary, the QM follows the opposite procedure; it explores nature by applying mathematical tools to basic conjectures and "principles"[164]; obviously is a valid approach, but a much more laborious course; the mathematical exploration that in almost a century has failed finding that material particles are rings of oscillatory substance; while the TH by very simple logic assumptions, finds consistent explanations of the origin of the magnetic moment, energy-mass relation, fractional Spin, etc., all this in just the first chapters of its original proposal.

---

[164] Suppose you have to describe mathematically a rotating screw of a given length which is coming into a wood wall; you will have to simplify the equation, and describe the phenomenon as a sinusoidal wave of finite duration which is changing phase. The procedure will introduce systematic distortions because the model must use ideal mathematical objects, to describe an imperfect and finite natural entity, thus we will be obliged to describe the screw as a composition of many waves of diverse frequencies, and infinite duration. We may apply a Fourier analysis and picture this object it in the frequency domain; later we may go backwards and discover that the precise location of the screw cannot be found anymore. Similar procedures may arrive to unreal results because after applying mathematical operators (transformers, switches, commuters, etc.) to this imperfect description of a finite object; not because mathematics failed, but because the initial description was improper or incomplete.

The following analysis compares some of the basic principles of QM with the TH proposals.

## The uncertainty principle

The "uncertainty principle" is the mathematical statement of an indeterminacy that is fruit of the imprecise question applied to the terms "position" and "energy" of finite objects. Define the "position" of a traveling wave of certain length, and at a given time, you must first specify to which part of the wave the question refers to, because obviously each part of it occupies a different position at a given moment; on the other hand, in order to know the contained energy one must know the "size" of the particle because it defines the content of matter/energy; therefore, measure more precisely the energy, one needs more time to do it, consequently decreases the accuracy of the precise position at a given time.

This uncertainty is the result of the implicit vagueness of the question, arising from the fact that energy (mass or time) implies measuring duration, and position refers to an instant value; the faster you quantify, the less accurate is the energetic evaluation and vice versa. Heisenberg's proposal states that the product of the standard deviation of the position ($\sigma x$) and standard deviation of the moment ($\sigma p$) is a constant, proportional to the reduced Planck constant ($\sigma x\, \sigma p = h_r / 2$); this is a statistical manner of expressing this principle. On the contrary, the TH expresses the same reality with other parameters, with the equation ($I / T* = 2\, h_r$), being 'I' the inertia of the particle and 'T*' its period, this relation prevents using statistical parameters.

In summary, the TH starts proposing that every particle must be an oscillating physical entity with defined dimension; a moving object which can be precisely described and located in an environment of four dimensional time. The QM uses the opposite approach; initiates the analysis proposing principles and conjectures, which pose a statistical description of particle properties.

## The Schrödinger equation

When the QM considers the particles as material objects, it does not specify physical dimensions, and when considered as waves, sometimes treats them as a cord inside a rigid enclosure. The TH always considers particles as entities containing an oscillating substance, whose size is directly related to the period of this oscillation; not as "standing wave" vibrating at a "resonant frequency."

The Hyle equation "H($\tau$; t)" <u>describes the substance constituting any particle</u> as a three dimensional periodic vector function, with magnitude equal to one within the limits of its structure; it also shows that material particles are circular static entities while photons are linear structures travelling away from its source while rotating on a two-dimension plane. By following the most fundamental natural law, the "Law of Symmetry", the TH shows how a geometric analysis of this proposal explains the origin of the basic variables in physics and most of the observed natural phenomena.

The Schrödinger "Wave Function" "$\Psi$(r; t)" describes particles as oscillatory abstract entities in accordance to the paradigm of the "principle of wave-matter duality" proposed by De Broglie.

Both proposals may look similar because their equations look alike; include the same variables, and deal with the same object basically described by the Einstein-Planck equation. But there is a substantial difference: the TH is a proposal which describes how universe is constructed under a unique perspective; that energy is a secondary parameter derived from the elastic condition of Hyle structures; while the Hamiltonians and Laplacians used by the Schrodinger equations are equations applied to a Newtonian mechanics definitions, not contributing with any substantial explanation about the origin of this natural entity.

The TH proposes and demonstrates that a particle is not a dual abstract entity, it is a substantial entity always occupying a location relative to other entities; it has geometrical and physical properties, and it interacts ruled by precise geometric theorems.

The standing wave proposed by Schrodinger, could be associated to the way the TH describes the symmetric condition of any stable particle as a <u>static and unchanging three dimensional entity;</u> it may seem similar to, but is not a standing wave.

The mathematical tools and procedures followed by Schrodinger are valid and could be applied to Hyle structures, and obtain similar results. From a mathematic perspective, the differences between both visions may appear being merely formal, but there is a profound difference from the physics point of view, because in the TH the proposed entity is engendered as the substantial building block ruled by a single law "the law of symmetry" which explains coherently diverse phenomena as gravity, force, elasticity, etc., "rediscovering" fundamental physics laws/equations and finally explaining with the same components the complex atomic mechanics. On the other hand, the "wave function" describes a vague mathematical entity with uncertain geometry and undetermined location; all that it describes accurately is the oscillating energy in a particle.

## De Broglie's hypothesis

De Broglie proposed a relation between the wavelength '$\Lambda o$' and the momentum '$p_o$' of the particle, as follows:

$$\Lambda o = h / m v = h / p_o$$

That is:
$$p_o = h_r \, \Omega o$$

Where 'h' is the Planck constant and the angular velocity ($\Omega o = 2 \pi / \Lambda o$) is not the angular velocity '$\omega$' of Hyle, but an arbitrarily assigned angular frequency, or wavelength, function of the mass 'm' and momentum '$p_o$' of the particle when regarded as a wave.

By linking a wavelength ($\Lambda o$) and momentum ($p_o = m\,v$) the proposal takes for granted that the particle is a wave; the relation ($p_o = h_r\,\Omega o$) indirectly assigns to '$h_r$' the quality of a moment of inertia ($L = I\,\Omega o$) as if the particle would be a mass rotating along a circle of perimeter '$\Lambda o$'. These relations establish a dogmatic "principle": that sometimes particles act as waves and sometimes as matter; giving rise to the concept of "wave-matter duality."

In contrast, the TH proposes that a particle is an oscillating substance (the Hyle) travelling on a circular path forming a static ring with radius ($r = c\,T / 2\,\pi$) and perimeter ($\lambda = c\,T$) whose mass is ($m = h\,f / c^2$); therefore, for the TH Hyle is not "mass", and the tangential speed is 'c'; not 'v'.

The QM assumes that the double-slit experiment proves the wave-matter attribute of particles, because an electron beam passing through them, concentrates in some specific directions, as does a "light wave." This assessment includes several assumptions: that "light" is an oscillating substance, that photons are similar to material particles, and that there is no other interpretation for the experimental observations.

According to the TH, the double slit experiment results in particle concentration because the randomly oriented particle rings collide with the walls of the slot bouncing, and then follow some preferred trajectories. Similar results are observed in the Gerlach experiment where the electrons are previously oriented by an heterogeneous magnetic field, separating the rings rotating in opposite directions in two separate paths, so they concentrate in two groups.

## Spin

The QM considers the "spin" as a property of material particles, related to its magnetic moment. The TH shows that the spin proves that particles are rings. Their angular momentum when they rotate around its diameter is one half of the angular momentum when they rotate around their center.

FF

# ANNEX II

## Table: New Natural System of Units

| PARTICLE PARAMETER | Symbol | [unit] |
|---|---|---|
| Compression {T* - T)/T} | $\xi$ | [] |
| Conversion factor ($\tau$ to d) | c | [] |
| Fine structure constant | $\alpha$ | [] |
| Hyle vector H($\tau$;t) | H | [] |
| Metrix ($\Delta\tau/\Delta t$) | $\mu$ | [] |
| Moment of Inertia (I = H) | I | [] |
| SPIN | s | [] |
| Unit electric charge | q | [] |
| Velocity ($\Delta\tau/\Delta t$) | v | [] |
| Arbitrary constant | kte | [any] |
| Relative time **($d\zeta = d\tau + dt$)** | $\zeta$ | [sec] |
| Photon period | $T_0{}^*$ | [sec] |
| Photon length | $T_0$ | [sec] |
| Particle period | $T_1{}^*$ | [sec] |
| Particle perimeter | $T_1$ | [sec] |
| Distance in TT | d | [sec] |
| Displacement direction | z | [sec] |
| Wave length ($\lambda = c\,T_0$) | $\lambda$ | [sec] |
| Sub-space G constant | $A_g$ | [sec] |
| Sub-space E constant | $A_e$ | [sec] |
| Absolute time (t) | t | [sec] |
| Travel time ($\tau$) | $\tau$ | [sec] |
| Gravitational constant | G | $[sec]^3$ |
| Gravitational factor | $K_g$ | $[sec]^3$ |
| Electric field constant | ke | $[sec]^5$ |

| PARTICLE PARAMETER | Symbol | [unit] |
|---|---|---|
| Unit vector in dimension x | $x^o$ | 1 [X] |
| Radial angular momentum | Lr | 1/[sec] |
| Axial angular momentum | Ld | 1/[sec] |
| Planck constant | h | 1/[sec] |
| Angular velocity | $\omega$ | 1/[sec] |
| Acceleration | a | 1/[sec] |
| Curvature | $\kappa$ | 1/[sec] |
| Magnetic field | B | 1/[sec] |
| Energy | $\varepsilon$ | $1/[sec]^2$ |
| Mass | m | $1/[sec]^2$ |
| Force | F | $1/[sec]^3$ |
| Elastic constant | k' | $1/[sec]^4$ |

A

B

CPSIA information can be obtained
at www.ICGtesting.com
Printed in the USA
FFOW02n1624110816
26578FF